9 Klu

PEANO

STUDIES IN THE HISTORY OF MODERN SCIENCE

Editors:

ROBERT S. COHEN, *Boston University*

ERWIN N. HIEBERT, *Harvard University*

EVERETT I. MENDELSOHN, *Harvard University*

VOLUME 4

GIUSEPPE PEANO (1858 – 1932)

HUBERT C. KENNEDY

Department of Mathematics, Providence College, Providence

PEANO

Life and Works of Giuseppe Peano

D. REIDEL PUBLISHING COMPANY

DORDRECHT : HOLLAND / BOSTON : U.S.A.

LONDON : ENGLAND

Library of Congress Cataloging in Publication Data

Kennedy, Hubert C.
    Peano, life and works of Giuseppe Peano.

    (Studies in the history of modern science ; v. 4)
    "Chronological list of the publications of Giuseppe Peano": p.
    Bibliography: p.
    Includes indexes.
    1.   Peano, Giuseppe, 1858–1932.   2.   Mathematicians—Italy
—Biography.   I.   Title.   II.   Series.
QA29.P42K47   510′.92′4 [B]   80–10880
ISBN 90–277–1067–8
ISBN 90–277–1068–6 pbk.

Published by D. Reidel Publishing Company,
P.O. Box 17, 3300 AA Dordrecht, Holland

Sold and distributed in the U.S.A. and Canada
by Kluwer Boston Inc., Lincoln Building,
160 Old Derby Street, Hingham, MA 02043, U.S.A.

In all other countries, sold and distributed
by Kluwer Academic Publishers Group,
P.O. Box 322, 3300 AH Dordrecht, Holland

D. Reidel Publishing Company is a member of the Kluwer Group

Printed in The Netherlands

In Memory of
Kenneth O. May
(1915—1977)

PREFACE

All students of mathematics know of Peano's postulates for the natural numbers and his famous space-filling curve, yet their knowledge often stops there. Part of the reason is that there has not until now been a full-scale study of his life and works. This must surely be surprising, when one realizes the length of his academic career (over 50 years) and the extent of his publications (over 200) in a wide variety of fields, many of which had immediate and long-term effects on the development of modern mathematics. A study of his life seems long overdue.

It appeared to me that the most likely person to write a biography of Peano would be his devoted disciple Ugo Cassina, with whom I studied at the University of Milan in 1957–58. I wrote to Professor Cassina on 29 October, 1963, inquiring if he planned to write the biography, and I offered him my assistance, since I hoped to return to Italy for a year. He replied on 28 November, 1963, suggesting that we collaborate, meaning by this that I would write the biography, in English, using his material and advice. I gladly agreed to this suggestion, but work on the project had hardly begun when Professor Cassina died unexpectedly on 5 October, 1964. I then decided to continue the project on my own.

I spent the academic year 1966–67 in Turin; completion of the book took ten years. In the meantime, however, I gave priority to translating a selection of Peano's writings, which was published as *Selected Works of Giuseppe Peano* (University of Toronto Press, 1973). Fortunately, I have been able to gather some information from people who knew Peano personally, but much documentary evidence has been lost, and it continues to disappear. (Some material I have seen has since disappeared, even from places where I thought it secure. One librarian, who had been given a number of Peano's postcards, destroyed them with the comment that they had no place in a 'library'.)

Consequently, I gave more attention to Peano's publications. To my surprise, I was able to increase Cassina's already extensive list by about 20%. Because many of these publications are rare and difficult to find, I have mentioned and described the contents of each, using for reference the numerical ordering begun by Peano himself and continued by Cassina. These references are placed within brackets.

ix

A large number of people and institutions have helped in the preparation of this biography, and I am grateful to them all. My particular thanks go to Peano's nieces Carola, Maria, and Caterina, who shared their recollection of their uncle with me and furnished the photograph that appears as frontispiece. I would not have begun the biography without the suggestion of Ugo Cassina and would not have completed it without the encouragement of Kenneth O. May, who called it a "priority project in the history of mathematics." Among institutions whose libraries and librarians have been of assistance should be mentioned: Akademie der Wissenschaften der DDR (Berlin), Bertrand Russell Archives (McMaster University, Hamilton, Ontario), Biblioteca Civica di Cuneo, Brown University (Providence, Rhode Island), Columbia University (New York), Eidgenössische Technische Hochschule (Zürich), Institut Mittag-Leffler (Djursholm, Sweden), Niedersächsische Staats- und Universitäts-bibliothek (Göttingen), Staatsbibliothek Preussischer Kulturbesitz (West Berlin), and Università di Torino.

I acknowledge my debt of gratitude to all these and many others not mentioned. I alone am responsible for this work's many deficiences, of which I am painfully aware.

HUBERT KENNEDY
San Francisco, 11 July 1979

TABLE OF CONTENTS

Preface   ix

1. Childhood   1
2. Student Years   3
3. University Assistant   7
4. Making a Career   20
5. Extraordinary Professor   35
6. The *Formulario* Project   44
7. Ordinary Professor   51
8. The Controversy With Volterra   56
9. The First International Congress of Mathematicians   64
10. Contact With Frege   73
11. Peano Acquires a Printing Press   78
12. The School of Peano   84
13. Paris, 1900   91
14. The Decline Begins   100
15. Latino Sine Flexione   107
16. The Cotton Workers' Strike   114
17. Completion of the *Formulario*   118
18. Academia pro Interlingua   125
19. Apostle of Interlingua   136
20. The War Years   144
21. The Postwar Years   152
22. The Toronto Congress   159
23. The Final Years   165
24. Afterwards   169
25. Summing Up   172

Appendix 1. Peano's Professors   176
Appendix 2. Members of the School of Peano   187
Appendix 3. List of Papers by Other Authors Presented by Peano to
  the Academy of Sciences of Turin   188

Chronological List of the Publications of Giuseppe Peano   195

Bibliography   211

Index of the Publications of Giuseppe Peano   217

Index of Names   221

Index of Subjects   229

CHAPTER 1

CHILDHOOD

Cuneo, capital city of the province to which it gives its name, is located in northeastern Italy, halfway between the industrial center of Turin to the north and the city-state of Monaco on the Mediterranean coast to the south. On the east side of Cuneo flows the stream called 'Torrente Gesso' and on the west flows the Stura River. Their confluence forms the 'wedge' (*cuneo* in Italian) that gives the city its name. Across the Torrente Gesso, about three miles from Cuneo, is the crossroads village of Spinetta, and nearby is the farmhouse called 'Tetto Galant', where Giuseppe Peano was born on 27 August, 1858. The name is typical of the region, if more elegant than others, such as 'Tetto dei Galli' or 'Tetto del Lupo'. The entrance to the living quarters is through the kitchen, which is almost on ground level. Up a few steps is the living room and opening from this is the bedroom, which gives onto a balcony with steps leading down to the garden below, through which a small brook flows. A steep stairway in the living room leads into the attic, where Giuseppe slept as a child.

He was the second child of Bartolomeo and Rosa Cavallo Peano. His brother Michele was seven years older than he; there were two younger brothers, Francesco and Bartolomeo, and a sister, Rosa. He first attended the village school in Spinetta, each day in winter carrying the stick of firewood that the pupils were asked to bring for the school stove. He next joined his brother in the daily walk to school in Cuneo. Later their parents relieved them of this by renting their home and field, and moving into Cuneo. There they lived in two rooms in the area Lazzareto, later named Baluardi Gesso. When Francesco and Rosa completed elementary school, they returned to Tetto Galant to work in the field with their father. The other children stayed in Cuneo with their mother.

The oldest son, Michele, went on to technical school and became a sucessful surveyor. He married Giuseppina Pellegrino and had seven children: Michele, Alessio, Carmelo, Giuseppina, Carola, Caterina, and Maria. Francesco married Lucia Giraldo, Bartolomeo became a priest, and Rosa married Luigi Bernardi.

Giuseppe's uncle (mother's brother) G. Michele Cavallo was a priest and lawyer in the episcopal curia in Turin. Seeing that Giuseppe was ahead of his

1

fellow pupils, this uncle invited him to Turin to complete his studies. Thus Giuseppe left Cuneo sometime in 1870 or 1871 and lived with his uncle in Turin, although his summer holidays were spent helping his parents in their work on the farm. In Turin he received private instruction at the Istituto Ferraris and from his uncle, taking final exams at the Ginnasio Cavour in 1873. He then enrolled as a pupil in the Liceo Cavour, where he received the *licenza liceale* in 1876.

Giuseppe's ability was early recognized, prompting his uncle to invite him to Turin. But the sacrifices of his parents for the education of their children must be recognized. Their dedication was extraordinary in a time of widespread rural illiteracy. His background was later slightingly referred to as 'sub-proletarian', but Giuseppe was proud of his origins and retained his love of the land. Many anecdotes are told to illustrate this. For example, once while a university student, during the summer, he went with his father into Cuneo to buy some sheep at the market, and he drove them home while his father stayed on other business. Later in life he was happiest during the time spent at his villa in Cavoretto near Turin. There he bought a piece of land that he always called his 'Eden'. He retained love of animals, too, especially dogs, giving them names from Greek mythology: Argo (Argus), Febo (Phoebus Apollo), and Melampo (Melampus).

Peano did not continue practicing the Catholic religion in which he had been reared, despite the influence of his uncle and his continued affection for his brother Bartolomeo, who became a priest. Typically, when later in life he took a niece or nephew on a trip with him, he would take them to a church on Sunday and return for them after the Mass was over.

Giuseppe loved learning; his final exam at the Liceo Cavour was very good and he obtained a room-and-board scholarship at the Collegio delle Provincie, that had been established to make university study possible for young people from the provinces. Thus began an association with the University of Turin that was to last until his death 56 years later.

CHAPTER 2

STUDENT YEARS

Peano officially enrolled as a student at the University of Turin on 2 October, 1876, when he duly paid the 40 lire enrollment fee and one-half of the year's tuition fee of 132 lire. (He paid the second half on 20 April, 1877.) It was probably after then that he moved into the Collegio delle Provincie. The Solemn Inauguration of the academic year took place in the Great Hall of the University on Monday, 20 November 1876. The rector was Ilario Filiberto Peteri, professor of cannon law. He had been rector since 1874 and this was his last year. The principal speaker was Luigi Schiaparelli, professor of ancient history, who spoke 'On the latest advances in the history of the ancient Orient and their relation to the future of the Royal University of Turin.' Peano may have heard Schiaparelli's talk. It is more likely that he was busy getting acquainted with his classmates at the Collegio delle Provincie, six of whom were also among the 103 students who registered for the first year of the lower two-year program in the mathematical and physical sciences. These six were: Michele Barale, Emilio Favre, Luigi Foresti, Carlo Losio, Raimondo Miracca, and Giuseppe Signorelli. All but Barale and Foresti completed the two-year program in 1878, but only Peano continued in mathematics. Most of the others entered the engineering program, as Peano also had originally intended to do.

Classes began the next day. Peano's first class was with Enrico D'Ovidio in analytic geometry. This class met at 8:45 in Room 17. Peano's next class was at 12:30 in Room 16b: ornamental design with Carlo Ceppi. At 2:00 he returned to Room 17 for a class in projective geometry and design. Giuseppe Bruno was listed as the professor for this course, but it was actually taught by Donato Levi. On Wednesday Peano's first class was at 10:15 in Room 20. The subject was chemistry and the professor Hugo Schiff (or Ugo Schiff, as he was known in Turin). Design was Peano's only other class on Wednesday. On Thursday at 8:45 in Room 17 he had his first class in algebra. The professor was D'Ovidio again. Thus Peano had five oourses that year: algebra, with classes on Monday and Thursday; analytic geometry, with classes on Tuesday and Saturday; projective geometry, with classes on Tuesday, Thursday, and Saturday; chemistry, with classes on Monday, Wednesday, Friday, and Saturday; and design, with classes every day except Thursday. Although

3

D'Ovidio's classes each met only twice a week, they were 1½ hours long, as were Ceppi's.

During the year 1876–77 there was a total of 1334 enrolled students at the university (and 404 *furori corso*, who still had exams to take), of which 458 were first-year students, and of these 103 were enrolled in mathematics.

Near the end of Peano's first year at the university he competed for the prizes available to students in the Facultá di Scienze. The result of this competition was announced on 8 June 1877. The cash awards included two 'Balbo' awards and three 'Bricco e Martini' awards. Seven students were given honorable mention. Peano received fourth honorable mention and was the only first-year mathematics student in the list of awards. Already on 20 April, 1877, he had paid the examination fee. On Tuesday, 3 July, he had two exams: one was in chemistry and he received 8 of a possible 9 points; the other was in algebra and analytic geometry and he received 9 of 9 'with honors'. On the following Tuesday he had an exam in projective geometry and received again 9 of 9.

Peano probably left shortly afterwards to spend the summer with his family in Cuneo. He was back in Turin by Monday, 19 November, 1877, when classes began for the new academic year. (He paid the first half of his tuition fee on that date.) Except for Ceppi, he had all new professors that year.

Probably Peano arrived in Turin in time to attend the Solemn Inauguration of the academic year on Saturday, 17 November. The principal speaker was Michele Lessona, professor of zoology and rector of the university. Peano had elected to take the zoology course with him that year. Lessona spoke 'On the zoological studies in Piedmont.'

Peano's first class of the new academic year was at 8:00 in Room 21. This was the zoology course with Lessona. It met every day. His second class was at 9:00, again in Room 21. This was mineralogy and geology with Giorgio Spezia, a course intended for those planning to continue in engineering. It met on Monday, Wednesday, and Friday. At 10:15 Peano went to Room 17c for the course in infinitesimal calculus which met on Monday, Tuesday, Thursday, and Saturday. The professor was Angelo Genocchi, although Eligio Martini seems to have shared some of the responsibility. His design class with Ceppi met (every day except Thursday) at 12:30 in Room 15. At 2:00 Peano had a physics class in Room 9 with Giuseppe Basso. This class met on Monday, Wednesday, and Friday. Thus Monday was a very full day of classes for Peano since all of his courses, except descriptive geometry met on that day. Descriptive geometry was taught by Giuseppe Bruno and met on Tuesday, Thursday, and Saturday at 2:00 in Room 17b.

By the end of Peano's second year there were 106 students enrolled for the second-year concentration in pure mathematics. Peano again competed for the 'Balbo' and 'Bricco e Martini' prizes. This time he placed fourth in the list of prize winners, the only second-year student to receive one of the six money awards. Peano's award of 200 lire more than made up for the tuition fee of 132 lire (which he had again paid in two installments) and the exam fee of 20 lire.

Peano took his exams in July, earning the maximum number of points, nine, in each of the subjects: calculus, descriptive geometry, and physics. Only these exams are listed in the university register, where there is also a notation that a diploma of *licenza* was issued on 3 November, 1878 (a Sunday!) On the next day the inaugural lecture of the new academic year was given by Count Carlo Bon-Compagni, professor of constitutional law, on the topic 'Ancient oriental despotism and the liberty of Greece.' Peano's third-year classes began the day following this.

At first Peano had planned to continue in the Royal Engineering School, as indeed almost all his classmates did. He obtained there an exemption from the tuition fee. But he decided to remain in pure mathematics and was able, on 26 October, 1878, to have his fee exemption validated by the rector of the university. By 20 December, 1878, the university counted 1509 enrolled students and 608 *fuori corso*, but Peano found himself the only third-year student enrolled in pure mathematics!

According to the university yearbook, Peano's first class that Tuesday was geodesy with Giuseppe Lantelme, who was officially substituting for Camillo Ferrati. This class met Tuesday, Thursday, and Saturday, at 8:30 in Room 17b. This was followed by rational mechanics with G.B. Erba in the same room at 10:00, this class meeting on Monday, Tuesday, Wednesday, Friday, and Saturday. In the afternoon he had a class with Francesco Faà di Bruno in higher analysis. This class met at 2:00 on Tuesday, Thursday, and Saturday in Room 17c. All of these were new professors for Peano. He was already acquainted with Enrico D'Ovidio, whose class in higher geometry met on Wednesday and Friday at 8:45 in Room 17c. The university register has no record of any exams in these courses, but there is a notation, dated 13 June, 1879, that he was dispensed from the exam fee.

During his summer vacations, Peano rejoined his family on the farm Tetto Galant. He was very fond of his family and in later years helped them to the extent he was financially able.

Peano's fourth and final year as a student began officially on 3 November, 1879, when Domenico Tibone gave the inaugural lecture on 'The Role of Medicine in Modern Civilization.' D'Ovidio was now chairman of the Faculty of Mathematical, Physical, and Natural Sciences. Peano was again dispensed from all fees (from the tuition fee on 29 November, 1879, and from the exam and diploma fees on 1 June, 1880.) Classes began on Tuesday, 4 November. The regular courses for fourth-year mathematics students were astronomy, higher mechanics, and mathemathical physics. Peano did not take the course in astronomy, but added another course in higher geometry with D'Ovidio. Mathematical physics was taught by Giuseppe Basso. Only Francesco Siacci, who taught the course in higher mechanics, was new to Peano.

His final examination was held on Friday, 16 July, 1880. According to the university register, he was tested in rational mechanics, theoretical geodesy, higher geometry, higher mechanics, and mathematical physics. Four other students were examined that year. Only Peano and Antonio Barberis were successful. Barberis graduated with 'honors' (*pieni voti legali*) and Peano with 'high honors' (*pieni voti assoluti*). No one received 'highest honors' (*lode*). Peano received the maximum number of points (18 out of 18) and was duly proclaimed *dottore in matematica*. His diploma to that effect was issued on 29 September, 1880.

CHAPTER 3

UNIVERSITY ASSISTANT

Enrico D'Ovidio had shortly before been instrumental in persuading the Minister of Public Instruction to establish the position of university assistant. This gave young graduates a way of continuing their education and furnished a first step toward a university carrer. D'Ovidio, himself, always tried when possible to confer this position on the best graduates, which made the position of assistant for algebra and analytic geometry especially desirable. D'Ovidio's first assistant, Fortunato Maglioli, was appointed for the year 1878–79. He was followed in 1879–80 by Francesco Gerbaldi. Peano was appointed for the year 1880–81.

One of the 92 freshmen students under Peano's care during his first year as assistant was Giovanni Vailati, who would later be Peano's assistant. We may learn something of Peano's duties as assistant from the description of his own duties given by Vailati in a letter to his cousin Orazio Premoli on 22 December, 1892:

In the days just passed my *teaching* duties have left me little free time, not so much because of the frequency of the hours of school (nine a week, of which three consist of merely attending the lesson of the *principal*) as for the necessary preparation and elaboration of the topics to discuss so as to avoid the danger of pulling a *boner*.

For Peano, the hours of attending the lessons of the 'principal' were probably four, as D'Ovidio had algebra classes scheduled for Tuesday and Thursday, and analytic geometry classes scheduled for Wednesday and Saturday, all at 8:45. Peano had now moved from the Collegio delle Provincie, at No. 15, Via Po, to a room at No. 50 of the same street, so he had no difficulty meeting the early classes. He gained a reputation as a good teacher (although the reputation suffered in later years.) This was in spite of a somewhat hoarse voice, said to have been the result of an attack of pneumonia, and an inability to pronounce the letter 'r'. His students made fun of this, calling him "the professor of the *tle tle*", because of the way he said *tre* (three).

This year saw the first of Peano's over 200 publications. His first paper was read by D'Ovidio at a session of the Academy of Sciences (mathematical and physical sciences section) on Sunday, 10 April, 1881. This and the three publications of the following academic year, while containing some interesting ideas, were not very important for Peano's later development, since they only followed up some of the researches of his teachers, D'Ovidio and Faà di Bruno. (See in the Chronological List of publications [1–4].)

This year, too, Peano probably had increasing contact with Enrico Novarese, who was almost exactly Peano's age, and was one of only two awarded the *laurea* in mathematics at the close of the year. Novarese, who died prematurely in 1892, was one of the founders, with Peano, of the *Rivista di Matematica*.

The following academic year, 1881–82, Novarese became D'Ovidio's assistant and Peano became assistant to Angelo Genocchi, who held the chair of infinitesimal calculus. Thus Peano moved up with his students, calculus being a second-year course, and would again have had contact with Vailati. Classes were scheduled for Monday, Tuesday, Thursday, and Saturday, with exercise sessions on Wednesday and Friday, all at 10:15. Peano was certainly in charge of the exercise sessions.

This year was notable in the life of the university because of the student disturbances. The previous year D'Ovidio had been named rector of the university and began a new policy of reviewing in the yearbook 'the most notable things' of the preceding year. Of the year 1881–92 he wrote:

Of the discipline in the schools and the comportment of the students in general, no mention was made in my review of the preceding year. This was surely a good sign, for fortunate is the school about whose discipline there is nothing to say. But unfortunately I cannot entirely keep silent about this past year; and besides I am concerned that those far from the university campus might form exaggerated ideas about the import of anomalies which set off that year. I shall touch in a few words on how the agitations of the extremist parties, which preceded the last general elections, found an inopportune echo in a few students, who, making themselves the unconscious instruments of outside agitatiors, and profiting by the usual indifference of the many who neither associated themselves with nor opposed the rumors, through various pretexts provoked meetings and demonstrations in the halls of the university and through the streets of the city. At one time there was a dispute over the banner of the students, then a carnival trick carried out too long, then an anticlerical protest, and so on – a superficial ferment, a passing commotion. But in the meantime these harm the seriousness of study, the good result on exams, and the reputation of the entire student body. They distress the families, often enduring hardships in order to educate and instruct these young people. They put to the test the affection of the rector and the professors, who would rather proudly point out the youth committed to their care and dedicate themselves with no other distractions to complete the education of the heart as well as the mind. They set on fire the less studious and more rowdy students so as to disturb the tranquility in some schools at the risk of calling down rigorous measures on a whole course. And, what is worse, they induce in the young the very bad habit of going off into cries and uproars, feeding their vanity on easy triumphs with the mobs, as if Italy were not a fact and its completion were not to be expected from the works of those whose character is calm and serious and capable of quiet sacrifices.

The sentiments which I express here are not peculiar to me and my colleagues; they are shared by the great majority of our students, who on the solemn commemoration of Giuseppe Garibaldi and on other occasions have given ample proof, showing that they understand that it is up to the majority to uphold the decorum of the entire class. Since this is a great deal better than what would result through the fault of a very few worthless people and a few overexcited, who would lose nothing now by thinking exclusively of studying.

May they listen to the counsel of one who loves them, and precisely for that reason does not flatter them in a vulgar search for popularity.

In the spring of 1882 Peano made the first of a long series of discoveries that influenced the development and teaching of calculus. Genocchi was 65 years old and his health was beginning to fail. On 22 April, 1882, he was forced to interrupt his lessons. (He did not return to the classroom until 11 March, 1884.) Peano took over Genocchi's classes and, having to explain the theory of curved surfaces, he discovered an error in the commonly accepted definition of J.A. Serret (in his *Calcul intégral*, 1st edition, 1868) of the area of a curved surface. Peano gave a corrected definition in his lesson on 22 May. He recognized the importance of his discovery and lost no time in telling Genocchi about it. We may imagine his surprise on learning that Genocchi was already aware of the difficulty! In later years Peano showed little concern for priority, but he was doubtless disappointed at a time when he was anxious to launch a scientific career. But Genocchi had not discovered the difficulty; its discoverer was H.A. Schwarz.

As early as 20 December, 1880, Schwarz had written to Genocchi of his discovery and on 8 January, 1881, he wrote a brief account of his criticism of Serret's definition. Genocchi replied with a request that he write this up as a note for inclusion in the *Atti* of the Academy of Sciences of Turin, but Schwarz had not done this. When Genocchi learned of Peano's discovery, he immediately informed Schwarz of this (letter of 26 May, 1882.) Schwarz replied that he was having difficulties writing it up. On 28 July, Genocchi wrote to Schwarz: "I hope you will want to send it to me for the opening of the new academic year, and I urgently beg you to do so. For the subject is quite important, and I see that there are few people who suspect the faults of the usual definitions."

Genocchi had also written of the discovery of Schwarz and Peano to his friend Charles Hermite, who had also adopted the definition of Serret in his lessons. Hermite wrote to Schwarz asking for a copy of his criticism of Serret's definition so that he might include it in the forthcoming second edition of his course, and he wrote to Genocchi at the same time (22 September, 1882) to ask whether he should also cite Peano. Schwarz wrote the note requested and sent one copy to Hermite (for inclusion in the new edition of his course) and one to Genocchi for inclusion in the *Atti*. Hermite hurriedly included this in the early pages of his book and they appeared before the end of the year in lithograph, copies of which he sent to Genocchi and Schwarz. (The complete book was published in lithograph in the summer of 1883.) Naturally neither Schwarz nor Genocchi was pleased with this, since Schwarz did not like the appearance of his note in lithograph and neither he nor Genocchi wished to submit a note to the Academy of Sciences which had, in fact, already been published. Thus Schwarz' note was not *printed* until the second half of 1890, when it appeared in the second volume of his *Gesammelte mathematische Abhandlungen*. By that time Peano had published a note in the *Rendiconti* of the *Lincei* Academy (19 January 1890) [23], explaining his definition of the area of a curved surface using the notion of a bivector. (Peano notes there that the discovery of Schwarz was published in Hermite's book after the publication of Peano's lecture notes. He fails to mention the time sequence of the discovery!) Thus Peano has priority of publication, both in lithograph and in print, but there is no doubt that Schwarz made the discovery first.

Schwarz must have written to Peano immediately about this, for in a letter of 27 April, 1890, Peano wrote him: "Several days ago (the 12th of this month) I sent you a long letter in reply to your note on the erroneous definition of Mr. Serret." (The 12 April letter has apparently been lost.)

Peano had moved this year to No. 12, Via Milano, perhaps a twenty-minute walk from his classroom, but by the fall of 1882 he again had a room on Via Po, this time at No. 28, on the 4th floor. Tancredi Zeni is now D'Ovidio's assisttant and Novarese is assistant for rational mechanics. (Zeni, one of six graduates that year in mathematics, was the only one to graduate 'with honors'.)

Although Peano's schedule is listed in the university yearbook as the same as the year before, he was actually in complete charge of the calculus course. In fact, according to Genocchi's records, Peano substituted for him continuously from 22 April 1882 (Genocchi's last lecture) to 11 March, 1884 (Genocchi's first lecture.) Genocchi's long absence is explained by an accident that occurred while he was on holiday in Rivoli in September 1882. Because of his near blindness, he fell against a roadside guard post and broke a kneecap. As a result he had to remain immobilized for several months. After that he went out seldom, and did not return to the classroom for another year. He was 65 at the time of the accident.

In the spring of 1883, F. Siacci presented papers of Peano to the Academy of Sciences of Turin, at the sessions of 1 April and 20 May. The first of these [5] is his first publication in analysis. It contains, implicitly, his notion of integral, given explicitly in the *Calcolo Differenziale* of 1884. In it he introduces the concept of area (internal, external, and proper), using the notions of upper and lower bounds. This concept is analogous to those of length and volume which are explicitly defined in the *Applicazioni Geometriche* of 1887. The second note [6] deals with interpolating functions of a complex variable, and shows that certain of these may be expressed using an integral taken along a closed curve.

At the end of the academic year there were six graduates in mathematics, including Gino Loria (with high honors) and Corrado Segre (with highest honors.) Segre, like Peano, spent his entire academic career at the University of Turin, first as assistant to D'Ovidio and G. Bruno, and then, from 1888, as professor of higher geometry. Loria remained for three years as assistant to D'Ovidio, then transferred to the University of Genoa, where he became professor of algebra and analytic geometry.

By the fall of 1883 Peano had moved once more, to Via Po, No. 40. There is also a change in the calculus schedule. Instead of four hours of lectures and two of exercises, there are now three hours of each. As Peano was in charge of both, this probably made very little actual difference. In the meantime, however, an affair was developing that, although it led to a temporary break in the good relations between Genocchi and Peano, resulted in Peano's first major publication [8]. This book is something of a curiosity in the history of mathematics, since the title page gives the author as Genocchi, not Peano. The title is *Calcolo differenziale e principii di calcolo integrale* and the title page states that it is "published with additions by Dr. Giuseppe Peano." An explanation of the origin of this book is given in a letter that Peano wrote Genocchi on 7 June, 1883:

Esteemed professor,

A few days ago I was at Bocca book-publishers and the director, who is named Lerda, I believe, showed a great desire to publish part of a calculus text during the summer vacation, either written by you or according to your method of teaching: there is no need for me to add how useful such a work would be.

Would you please be so kind as to let me know whether it would be possible to firm up this matter in some way, i.e., if you do not wish to publish the text yourself, perhaps you might think it possible for me to write it, following your lessons, and if I did that, would you be willing to examine my manuscript before publication, or at any rate give me your valued suggestions and look over the proofs of the fascicles as they come off the press; or indeed, in the case that you do not wish to publish this yourself, it might not displease you if I went ahead and published the text, saying that it was *compiled according to your lessons*, or at least citing your name in the preface, because a great part of the treatment of the material would be yours, seeing that I learned it from you.

Please allow me, learned professor, to make a point of saying that I will do my best to see that all comes out well, and believe me

your most devoted student
G. Peano.

Genocchi's reception of this letter is expressed in a letter sent to his friend Judge Pietro Agnelli, of Piacenza, on 16 August 1883:

I could make a complete course in infinitesimal calculus, which I have lectured on for several years, but I do not have the strength to put everything in writing. It would *bore* me too much, to use our friend Marchese Mischi's usual expression, and seeing that Dr. Peano, my assistant and substitute, and also my former student, is willing to take on himself this trouble, I thought it best not to oppose it, and let him handle it as best he can.

After Genocchi's death (7 March 1889), Peano wrote a necrology for the university yearbook [21] and in it he gives his own version of the origins of this book:

He was many times urged to publish his calculus course, but he never did. Perhaps in the beginning this was because he had not yet reached the perfection he aimed at, and then in later years because he lacked the strength. It was only in '83 that he agreed to let the firm of Fratelli Bocca publish his lessons with my help, and this appeared in part the following year. He was ill at the time and wished to remain completely apart from the work. I used notes made by students at his lessons, comparing them point by point with all the principal calculus texts, as well as with original memoirs, thus taking into account the work of many. As a consequence, I made many additions and some changes in his lessons. Now, although these additions are for the most part distinguished either by my name or by being printed in smaller type, there are some, of less importance, which are not distinguished in any way. The result is that my publication does not exactly represent the professor's lessons. Nevertheless, as a text, it has received the approval of many scientists.

Fratelli Bocca announced publication for October 1883 of the first volume of Genocchi's 'Course in Infinitesimal Calculus', edited by Peano. When the publication did not appear on time, Felice Casorati, professor of infinitesimal analysis at the University of Pavia, wrote to Peano on 6 November asking about it. Peano replied in a postcard of 13 November:

Illustrious professor,
The calculus text that you ask about is now being printed. About 100 pages have already been printed, and some of these were placed at the disposal of the students, so that they could use them for the current year. The complete first volume can hardly

appear before the first of February, and at that time I will send you a copy without fail. If instead, you would prefer to have the fascicles as they come off the press, you have only to let me know.

Your devoted servant,
G. Peano.

Casorati replied on 2 December, declining Peano's offer, but thanking him for the information and saying that he wanted to know when the complete book would be ready so that he could recommend it to his students.

The same day that Peano wrote Casorati, Placido Tardy, professor at the University of Genoa, wrote Genocchi:

I had heard about the calculus course that Peano, following your lessons, proposed to publish, and I supposed that it would be revised by you. Now I hear that you do not wish to occupy yourself with it, but I am convinced that he will have your advice, and I hope to see the work as soon as it is published.

Despite the optimism of Fratelli Bocca and Peano, the complete work did not appear until the fall. (The preface is dated: Torino, 1 September, 1884.) Shortly thereafter disclaimers were sent to mathematical journals in Italy, France, and Belgium, and published over the signature of Genocchi. The Italian journal printed:

Recently the publishing firm of Fratelli Bocca published a volume entitled: Differential Calculus and Fundamentals of Integral Calculus. At the top of the title page is placed my name, and in the Preface it is stated that besides the course given by me at the University of Turin the volume contains important additions, some modifications, and various annotations, which are placed first. So that nothing will be attributed to me which is not mine, I must declare that I have had no part in the compilation of the aforementioned book and that everything is due to that outstanding young man Dr. Giuseppe Peano, whose name is signed to the Preface and Annotations.

The French journal said only: "I have had nothing to do with the preparation or publication of the work entitled: Differential Calculus and Fundamentals of Integral Calculus, published in Turin by the firm of Fratelli Bocca. This work is due exclusively to Dr. Peano." Paul Mansion, professor at the University of Ghent and editor of *Mathesis*, wrote Genocchi on 17 October and promised to insert Genocchi's disclaimer in the next issue of *Mathesis*. He also expressed curiosity about the book and asked Genocchi to lend him a copy for a week. Apprently Genocchi sent a copy to him, for Mansion wrote again on 28 October:

Sir and dear colleague,
    I have received the *Differential Calculus* published by Peano. It will be useful to me for several parts of my Course, which is just now going to press. Since you have had to

protest, I suppose that there are many things in it which were lifted from your lectures and that the book's merit, which is real, comes in great part from you. The November number of *Mathesis* will contain your letter.

Thanking you for your gift, I beg you to believe me, your devoted

P. Mansion.

It is easy to imagine that, reading one of these disclaimers, many people thought that Peano was using the name of Genocchi just to get his own research published, or at least was guilty of bad faith in dealing with his former teacher. Indeed, Charles Hermite, who had written to Genocchi on 6 October, 1884, congratulating him on its publication, wrote again on 31 October to commiserate with Genocchi on having such an indiscreet and unfaithful assistant. Genocchi complained to his friend Placido Tardy, who wrote to Genocchi on 8 November, 1884: "After your letter, I sent a visiting card as a thank-you note to Peano. I am surprised by what you tell me about the manner in which he has conducted himself toward you, in publishing your lessons." But it seems that Genocchi, who had a reputation for being quick-tempered, soon recovered and three weeks later Tardy could write: "I, too, am persuaded that Peano never suspected that he might be lacking in regard toward you."

In what way did Genocchi think Peano was "lacking in regard"? Ugo Cassina believed this was because of the words 'important additions,' written by Peano in the preface, and repeated by Genocchi in his disclaimer. He said: "In fact, we cannot deny that there is a certain youthful boldness in declaring 'important' the additions made to the lessons of the master, which in consequence lose somewhat in value." Cassina, who inspected Genocchi's unpublished manuscripts, also suggests that Genocchi may have been displeased also because of the "profound and essential difference between the text and his actual lessons."

Indeed, the additions *were* important, and the book has been much praised. Of the many notable things in this book, Alfred Pringsheim, in an article in the *Encyklopädie der mathematischen Wissenschaften*, cites: theorems and remarks on limits of indeterminate expressions, pointing out errors in the better texts then in use; a generalization of the mean-value theorem for derivatives; a theorem on uniform continuity of functions of several variables; theorems on the existence and differentiability of implicit functions; an example of a function whose partial derivatives do not commute; conditions for expressing a function of several variables with a Taylor's formula; a counter-example to the current theory of minima; and rules for integrating rational functions when roots of the denominator are not known. As a curiosity may also be mentioned the function which for rational x

equals 0, and for irrational x equals 1. P.G.L. Dirichlet had already considered this function in 1829, but Peano was the first to give an analytic expression for it, something that Gottlob Frege still believed in 1891 to be impossible.

Peano also included a new definition of the definite integral, using least upper bounds and greatest lower bounds of certain sums. He tells us that these l.u.b. and g.l.b. wre already introduced by V. Volterra and that he (Peano) had shown in 1883 [5] that if they are equal, then their common value is the Riemann integral, in the usual definition of a limit of Riemann sums. F.A. Medvedev, in discussing the development of the integral, says of this definition: "Peano took the last step in this [order of] thought by freeing the definition of the integral from the concept of limit" [Medvedev 1974, 214]. The book was translated into German by G. Bohlmann and A. Schepp in 1899, and there were two translations into Russian: one by N.S. Sineokov in 1903 and another by K.A. Posse in 1922.

Peano had discovered one of his strong points: the ability to uncover flaws in the teaching of calculus, to single out the difficulties in theorems and definitions (and to manufacture counter-examples simple enough to be understood and convincing), and to make theorems rigorous while simplifying their statements. The calculus text demonstrated all this, but in the meantime he sought to reach a more international, French-reading audience. He wrote a letter to the editor of the *Nouvelles Annales de Mathématiques* (published in the January 1884 issue) [7] criticizing the proof of an intermediate-value theorem for derivatives as given in Jordan's *Cours d'Analyse de l'Ecole Polytechnique* and Hoüel's *Cours de Calcul infinitesimal*, and he remarks that the mean-value formula: $f(x_0 + h) - f(x_0) = hf'(x_0 + \theta h)$, can be proved very easily without assuming the continuity of the derivative.

Before publishing this, the editor had sent a copy to Jordan, and Peano's published letter was followed by a brief extract from Jordan's reply, which begins. "I have no reply to the criticism of Dr. Peano, which is perfectly correct," and closes: "Mr. Peano says that it is easy to prove the formula $f(x + h) - f(x) = hf'(x + \theta h)$, without assuming the continuity of the derivative. I would be pleased to receive his proof, for I know of none which appears completely satisfactory to me."

Philip Gilbert, professor at the University of Louvain, came to Jordan's defense in a letter attempting to justify Jordan's proof of the intermediate-value theorem. With regard to Peano's remark that the mean-value formula can be proved without assuming continuity of the derivative, he says: "Mr Jordan asks, *non sans malice*, to see this proof, which is impossible, since the theorem is inexact," and he gives a counter-example. In his published reply,

Peano points out the error in Gilbert's attempted justification. He also points out that a theorem Gilbert used can be proved in a certain way, adding, with a touch of humor: "And I say this *sans malice*, for this way exists, but I shall leave the trouble of finding it to Mr. Gilbert." He then demolishes Gilbert's statement about the mean-value formula by furnishing a proof of his assertion, due to Ossian Bonnet, which he learned from Genocchi while a student, and he gives references to where correct proofs can be found in Italian and German texts of 1878, 1881, and 1882. (He sent a copy of the last proof, by Pasch, to Jordan.)

Gilbert had the last word, in the issue of October 1884. There has been a misunderstanding of terms, partly on the side of Peano, and partly on his side, but he does funish the proof requested by Peano, saying (*sans malice*?): "I have no need of the proposition in question, and that is why I have given little effort to perfecting the proof given below; but, since the theorem has interest in itself, I hope that Mr. Peano will want to publish his proof, which will without doubt be better."

In December 1884, Peano received the *privata docenza*, an event noted in his annual report in the university yearbook for 1884—85 by D'Ovidio, who had been rector of the university since 1880. This title meant that Peano was judged as qualified to be a university professor. By the fall of 1884 Peano had moved to Via Artisti, No. 31, still only a short walk from the university.

While Peano's career seemed to be advancing smoothly, the university was going through one of its many periods of student unrest. The evening of 11 March, 1885, the police arrested several students who were present at a demonstration protesting the occupation of Massawa (Eritrea) the month before. In the following days large numbers of their fellow students demanded their release. When this was not done, rioting broke out, both inside and outside the university. When the efforts of the rector and several professors to quiet the students had little effect, the Academic Council, with the approval of the government, decided to close the university on 15 March. It did not reopen until 15 April, a month later. Although the rector could report that fall that the conduct of the students had since been "worthy of all praise," the university regulations were modified as a result of a report to the king, that said in part:

We did not believe that university associations could be tolerated much longer, that were formed for political ends. These could not help but distract the students from their more immediate task, namely study, and disturb the peacefulness of our schools by bringing in passion, impatience, and commotion, all foreign to science.

In the fall of 1885, while continuing to teach calculus, Peano also substituted for Genocchi in a course titled 'Geometrical applications of infinitesimal calculus.' (Two years later he would publish a text with this title [11].) This course met three times a week. In addition, he taught, as *libero docente*, a course in 'Infinitesimal geometry, treated synthetically,' which met two times a week. He had now moved to rooms on the fourth floor of No. 25, Piazza Castello. This was his fifth move in six years, but he stayed at this address for four years.

In the spring of 1886 a note of Peano's was presented to the Academy of Sciences [9] in which he proved, for the first time, that the differential equation $y' = f(x, y)$ has a solution, on the sole assumption that f is continuous. The proof is elementary, if not entirely rigorous. (He gave a more rigorous proof of a generalization of this theorem in 1890 [27].) This theorem is typical of Peano's constant search for simplification. In a sense, it is the basic theorem of differential equations.

It is clear that f must satisfy some condition if the existence of a solution of $y' = f(x, y)$ is to be guaranteed. Early in the century Cauchy had shown that a unique solution exists if both f and its partial derivative with respect to y are continuous. Improvements on this theorem were made by C. Briot and J. Bouquet in 1856, and especially by R. Lipschitz in 1868. In a footnote to his article, Peano mentions these and others, closing with: "Mr. Vito Volterra (*Giornale di Matematiche*, Vol. 19) generalized their results, still leaving in doubt the truth of the theorem, supposing only the continuity of $f(x, y)$. In the present note we have also the answer to this question."

In the fall of 1886, Giorgio Anselmi, now rector of the university, could report: "I cannot fail to be happy over the fact that no disorders disturbed the tranquility of study in our University during the academic year 1885–86 just past. In fact, the conduct of the students throughout the whole year was worthy of all praise."

On 21 October, 1886, Peano was appointed professor at the Royal Miilitary Academy, adding the teaching of calculus there to his already heavy schedule at the university (a repeat of the previous year.) University professors often taught at the Military Academy, whose buildings were just across Via Verdi from the university. Peano would continue at the Military Academy for fifteen years.

By the spring of 1887 Peano had discovered the method of successive approximations for solving systems of linear differential equations. This was presented to the Academy of Sciences on 20 February [10]. (A modified version of this note, in French, was published the following year in the

Mathematische Annalen [12].) He always believed that he was the first to discover this method, and he tried several times to vindicate his priority over Emile Picard, who, Peano believed, began using it in 1891. In fact, Picard had already given an example of successive approximations in 1888 and at that time gave credit for the method to H.A. Schwarz.

This year also saw the publication by Fratelli Bocca of the text *Applicazioni geometriche del calcolo infinitesimale* [11], based on Peano's course at the university. He used the occasion to explain, in the preface, the circumstances of the publication of the *Calcolo differenziale* of 1884. He then quoted Genocchi's disclaimer and added: "Hence, though I made extensive use of the lessons of the illustrious professor, I assume the entire responsibility for what is written in that book, just as if no other name but mine appeared on the title page." He then thanks his friend Gino Loria "for the care he took in the correction of this book."

Of the many interesting things in this book, particular mention may be made of Peano's notion of the measure of a point set. Cantor had already considered this problem, but his definition of 'measure' was such that the measure of the union of two disjoint sets could be less than the sum of the measures of the two sets. Peano overcame this difficulty by defining, besides the Cantorian measure $m(A)$ of a point set A contained, say, in an interval I, its 'interior measure' $m(I) - m(I - A)$. He then called A measurable in case these two numbers coincide. This definition was also given the same year, and independently, by C. Jordan. Many familiar point sets, however, were not measurable by this definition, such as the rational numbers contained in a closed interval. Such sets also became measurable when H. Lebesque, in 1902, allowed the consideration of a countably infinite collection of sets, in place of the finite collection of sets used in Peano's definition. Peano's continuing interest in quadrature formulas also shows up here in the new formulas for the remainder terms in the trapezoidal formula and Simpson's formula.

Peano's friend Giulio Vivanti reviewed the book for the *Jahrbuch über die Fortschritte der Mathematik* and said, in part:

This work, which, to be sure, does not go beyond the elements of infinitesimal geometry, is of value for its originality of methods and its careful, exact exposition. The constant application of the segment-theory of Möbius–Grassmann, which brings with it a conciseness and elegance that can scarcely be found in quaternion theory, the exact foundation of the concept of limit in geometry, and finally the introduction of point sets and functions of these sets, by which infinitesimal geometry is brought to the same measure of rigor and generality that the infinitesimal calculus recently attained – all these characterize this unique book.

The functions of sets, mentioned by Vivanti, took up an earlier idea of Cauchy, that Peano extended to include the differentiation and integration of one function of a set by another. But this aspect of the work was largely ignored and by the time he returned to it in 1915 [175], it had been duplicated by H. Lebesque, W.H. Young, and others. Peano's pioneering efforts went unrecognized for a long time, but according to F.A. Medvedev:

The highest achievement of mathematicians of the 19th century in the working out of the theory of functions of sets is the fifth chapter of Peano's book. . . . Today it may be possible to find fault with regard to this or that method of reasoning applied by Peano, but at the same time, for the generality and depth of its various ideas, this chapter of Peano's book is more remarkable than Lebesque's memoir of 1910, that is generally considered as the source from which grew the boundless river of modern research in the theory of functions of sets [Medvedev 1975, 67–68].

CHAPTER 4

MAKING A CAREER

There seems little doubt that Peano had for some time aspired to the chair of infinitesimal calculus — and with reason. He had already proved to himself and his students that he was a capable teacher. In the seven years since his graduation he had published several articles and two quite remarkable books. He was *libero docente* and *assistente* at the university; indeed, in actuality he had taken over Genocchi's course. And he was a professor at the Military Academy. No doubt he felt his academic career reasonably secure.

On Thursday 21 July, 1887, Peano married Carola Crosio, the youngest of the four daughters of Luigi Crosio, a genre painter who specialized in Pompeiian and 17th century scenes. Carola and her sisters (Camilla, Aneta, and Bianca) all served as models for their father, who made pictures for lithographic reproduction by publishing firms in Milan, London, Zurich, and elsewhere. His paintings figured in several exhibitions of the Society for the Promotion of Fine Arts, which was founded in Turin in 1842. One of his oil paintings, 'The Curate's Bible', was owned by the Civic Museum of Turin and one of his paintings hung in the Church of San Giorgio, in the town of Chieri (Province of Turin.) Crosio was born in Alba in 1835 and died in Turin in 1915.

After a brief honeymoon trip to Genoa, Livorno, and Florence, the couple returned to Turin on 28 July. They lived at Peano's Piazza Castello address (No. 25) for two years, moving to the Corso del Valentino (No. 1, 3rd floor) in 1889.

On 10 October 1887 Peano wrote Felix Klein, saying that he had received a letter from Klein much earlier, but using his marriage as an excuse for not writing earlier. Peano wrote: "I would be glad to send the articles for publication in the *Mathematische Annalen*, making the additions that you suggest." Presumably, the reference is to the article, in French, treating the solution of differential equations by the method of successive approximations that was published in 1888 in the *Mathematische Annalen* [12]. This article is a modified version of a similar publication in Italian in 1887 [10].

Peano's schedule for the academic year 1887–88 was the same as before, except that he has now dropped the course in infinitesimal geometry. Anselmi, rector of the university, is once again happy to report no student disturbances during the previous year. Peano's father died during the course of this academic year, on 6 February, 1888.

In 1888, besides the modified French version of his note on successive approximations [12], Peano published a brief note on a geometrical definition of elliptic functions [13], which also had a Portuguese translation by F. Gomes Teixeira, and then the *Calcolo geometrico, secondo l'Ausdehnungslehre di H. Grassmann, preceduto dalle operazioni della logica deduttiva* (Geometrical calculus, according to the Ausdehnungslehre of H. Grassmann, preceded by the operations of deductive logic) [14]. The preface (dated 1 February, 1888) begins:

The geometrical calculus, in general, consists of a system of operations to be carried out on geometrical entities, analogous to those of algebra for numbers. It permits the expression by formulas of results of geometrical constructions, the representation by equations of geometrical propositions, and the substitutions of the transformation of equations for proofs. The geometrical calculus presents an analogy with analytic geometry; it differs from it in this, that while in analytic geometry calculations are made on the numbers that determine the geometrical entities, in this new science the calculations are made on those very entities.

The (unnumbered) chapter on the 'operations of deductive logic,' that opens the book is Peano's first publication in mathematical logic. It is based on his study of E. Schröder, G. Boole, C.S. Peirce, and others. Although he says that "several of the notations introduced here are used in the geometrical calculus," this part of the book actually has very little connection with what follows. His interest seems to be prompted by the analogy with the operations of algebra and geometrical calculus that is presented by the operations of mathematical logic. He says: "Deductive logic, which belongs to the mathematical sciences, has made little progress up to now. ... The few questions treated in this introduction constitute an organic whole, that can serve in many kinds of research." But what kind of research does he have in mind for mathematical logic? He gives two examples of "interesting" problems, one he has already solved: Given n classes, how many propositions may be stated using the logical symbols introduced in this book? He gives a general formula, for which, e. g. if $n = 2$, the number is 32766. An open question is: Given a relation between two variable entities, what are the classes and propositions that can be stated by using that relation and logical symbols?

In the preliminary chapter on logic, Peano develops first a calculus of classes and he uses the symbols $\cup$ and $\cap$ in their modern sense of union and intersection. All the entities of a system is designated by ⊘ and the empty set by ○. Thus AB = ○ is the statement that the intersection of A and B is empty, or in Peano's words, "no A is a B." The complement of A in the system is designated by $\bar{A}$. Thus $A\bar{B}$ = ○ is equivalent to "every A is a B", and for convenience Peano also introduces the symbolism $A < B$ to express this.

He next develops a parallel calculus of propositions using some of the same symbols. Here ○ means 'absurd' and ◉ means 'identical'. Negation of the proposition α is, however, expressed by $-\alpha$. Conjunction and alternation are expressed by ∩ and ∪, and $\alpha < \beta$ is read 'if α is true, then β is true.' No attempt is made to expound this axiomatically nor are there any proofs of the various formulas. Peano seems primarily concerned with showing that the propositions of logic can be treated in an algebraic fashion.

This publication gives only a hint of Peano's future development of mathematical logic. It should be seen as an attempt to synthesize the work of others. But his own systematic development of logic will begin almost immediately, with the introduction of new symbols (several used here are abandoned) and new concepts. Noteworthy here, however, is his recognition of an equivalence between the calculus of classes and the calculus of propositions.

The main part of the book, the geometrical calculus, is, by contrast, a polished work. It was already preceded by his *Applicazioni geometriche*, which used the notions of Möbius, Bellavitis, Hamilton, and Grassmann. He then decided that the methods of Grassmann were superior, this book resulting from his reworking of Grassmann's *Ausdehnungslehre*. Peano made no claim to originality for the ideas contained in it, but there can be no doubt that the extreme clarity of his presentation, in contrast with the notorious difficulty of reading Grassmann's work, helped to spread Grassmann's ideas and make them more popular.

Chapter I begins with definitions of 'lines, surfaces, and volumes.' The symbol AB is used for the line bounded by the two points A and B, and is thought of as described by a point P moving from A to B. (Thus AB is distinct from BA.) The measure of AB is gr AB (gr for *grandezza*.) The surface ABC is the triangle described by the line AP as P moves along BC from B to C. The tetrahedron described by the surface ABP as P moves long the line CD from C to D is the volume $ABCD$. Peano also denotes lines by lower case Latin letters, surfaces by lower case Greek letters, and volumes by upper case Greek letters. Thus, if a denotes AB, α denotes ABC, and A denotes ABCD, then the following are equivalent: A, αD, aCD, ABCD.

Letting Ω be an arbitrary, but fixed, volume and defining A/Ω to be the ratio of their measures, taken with + if they have the same direction and −

if they have opposite direction, Peano is able to compare various geometrical entities, e. g., $A = B$ if $A/\Omega = B/\Omega$, and establish an arithmetic in which sums of entities of one kind can even be 'multiplied' by sums of entities of another kind.

Peano's work was indeed readable, but it sparked the interest of only a few Italians, e.g. Burali-Forti and R. Marcolongo, whose long collaboration contributed much to the development of vector analysis in Italy. Peano, himself, stayed within the Grassmannian tradition and his writings contributed little further to the form of the emerging vector analysis. There can be no doubt, however, that his personal encouragement of younger mathematicians played an important role, and his interest in this topic continued. In the *Rivista di Matematica* of 1895 he announced the formation of the International Association for Promoting the Study of Quaternions and Allied Systems of Mathematics, and in 1901 he was National Secretary for Italy.

In the second half of the 20th century it is difficult to imagine the opposition to the introduction of vectorial methods. In fact, Burali-Forti was denied the *libera docenza* because of his insistence of using vectors, and this despite the efforts of Peano, a member of the judging committee, to persuade the other members of the committee to allow this. Despite this, Peano was certainly repaid in the coin requested in the preface of his book, where he wrote: "I shall be repaid for my labor in writing this book (and this is the only payment I expect) if it serves to spread among mathematicians some of the ideas of Grassmann." (Or was he complaining about his royalty arrangement with the publisher?)

Although it seems to have aroused no attention at the time, one of the most remarkable features of this book historically is that here, for the first time, is presented axioms for a vector space. This is given at the beginning of Chapter 9 and is worth quoting to show the clarity of Peano's ideas:

There exist systems of entities for which the following definitions are given:

(1) An *equivalence* of two entities of the system is defined, i. e. a proposition, designated by $a = b$, is defined that expresses a condition between two entities of the system, satisfied by certain pairs of the system and not by others, that satisfies the logical equations:

$$(a = b) = (b = a), (a = b) \cap (b = c) < (a = c).$$

(2) A *sum* of two entities a and b is defined, that is, there is defined an entity designated by $a + b$, that also pertains to the given system and that satisfies the conditions:

$(a = b) < (a + c = b + c), a + b = b + a, a + (b + c) = (a + b) + c,$
and the common value of the two members of the last equivalence is designated by $a + b + c$.

(3) Letting a be an entity of the system and m a positive whole number,
by the expression ma we shall mean the sum of m entities equal to **a**. It is
easy to see that, if **a**, **b**, ... are entities of the system and $m, n, \ldots$ are
positive whole numbers, then $(a = b) < (ma = mb)$; $m(a + b) = ma + mb$;
$(m + n)a = ma + na$; $m(na) = (mn)a$; $1a = a$.

We shall assume that a meaning is given to the expression ma, whatever
the real number m, so that the preceding equations are still satisfied. The
entity ma is called the *product* of the (real) number m by the entity **a**.

(4) Finally, we assume that there exists an entity of the system, that
we shall designate by 0, such that, whatever the entity **a**, the product of the
number 0 by the entity **a** is always the enitity 0, or

$$0a = 0.$$

If the meaning $a + (-1)b$ is given to the expression $a - b$, then it follows that:

$$a - a = 0, a + 0 = a.$$

Def. *Systems of entities for which the definitions* (1)–(4) *are given, so as to
satisfy the conditions imposed, are said to be* linear systems.

Peano then gives several examples of these 'linear systems' (vector spaces)
of various finite dimensions, mentioning the possibility of an infinite dimen-
sional vector space. In the 'Applications' at the end of the chapter he gives
as an example of an infinite-dimensional vector space the set of algebraic
entire functions of all degrees.

The basic ideas of this book were restated in a more succinct form six
years later, in a paper presented to the Academy of Sciences of Turin [90].

The Circolo Mathematico of Palermo published two observations of Peano
[14.1] concerning a note of F. Giudice that was presented at the sessions of
4 and 18 December of the previous year. In the first (dated 13 May, 1888)
Peano mentions Cauchy and gives a reference to his own *Calcolo differenziale*
of 1884. Giudice's reply follows, with several references to a work by Dini,
adding somewhat testily: "He [Peano] could recognize then, that it was not
at all necessary to remind me of his precious *Calcolo*: citing Cauchy, he could
very well think it superfluous to cite anyone later." His reply to the second
observation closes with the remark, "at most one could draw attention to
that in ordɔr to be exuberantly clear."

The matter of the argument is not very important in this case, but it
illustrates Peano's concern for rigor, and the reaction such people always

receive: "Do I have to state everything?" Peano was right, of course. Giudice's reply was especially unfortunate in that he quoted several sentences from Dini's work out of context, something that Peano would not pass up.

A second comment of Peano was published with the date 14 October, 1888, with the remark: "Seeing that the quotations contained in his reply could be misleading, I think it advantageous to point out the infidelity of some of them." The word "infidelity" hit the mark. Giudice replies that it is obvious that his omission was accidental, adding: "Nevertheless I thank the professor for pointing out the omission; but I reject the imputation of infidelity. Dini's book is too precious for anyone to want to touch it. He is not unfaithful, who gives precise directions to the original." (This exchange did not prevent a friendship between the two, and in fact Giudice became a firm adherent of the 'school' of Peano.)

This same issue of the *Rendiconti del Circolo Mathematico* contains a brief note of Peano, stating several geometrical theorems that can be proved using the methods of the geometrical calculus [15].

In 1888 Giovanni Vailati, called 'the Philosopher' by his classmates, graduated in mathematics. (He had already received an engineering degree in 1884.) He returned to his home in Lode, but visited Turin many times. In 1892 he returned to Turin as assistant to Peano for the calculus course at the university. In the meantime he became Peano's good friend and collaborator. In the fall, Mario Pieri, who had previously received an appointment at the Military Academy, became university assistant in projective geometry and Peano's good friend and collaborator.

The beginning of the academic year was marked throughout Italy by student disturbances. Anselmi reported:

The disturbances which occurred at the beginning of the current year in almost all the other universities of the Kingdom, occasioned by the monument to Giordano Bruno, to protest against the intrigues of the clerical party, also had an echo in our university, and we had a few disorders which, however, were of short duration and of no grave consequence.

The statue of Giordano Bruno, by Ettore Ferrari, was inaugurated in June of the following year, in the Campo dei Fiori in Rome.

Early in 1889 Peano published the remàrkable pamphlet (of 36 pages) *Arithmetices principia, nova methodo exposita* [16] that contains his first statement of the now famous postulates for the natural numbers. Dedekind had finished writing on 5 October, 1887, and published in 1888 his *Was sind und was sollen die Zahlen?*, which essentially contains Peano's analysis, but Peano did not see Dedekind's booklet until his own was going to press. *Arithmetices principia* is at once a landmark in the history of mathematical logic and of the foundations of mathematics. It is also something of a mystery. Why was it written in Latin? What is the significance of the title? The title translates: The principles of arithmetic, presented by a new method. The odd word in the title is *arithmetices*, which is not really Latin, but Greek. (The more truly Latin form would be *arithmeticae* or *arithmetica*.) Why did Peano choose this form — as a tribute to Euclid? Was the use of the word *principia* meant as a tribute to Newton? Peano was certainly aware of doing something historic.

But why Latin? If he had wanted to reach an international audience, he should have written in French, which he had already shown he was perfectly capable of doing. Nor does it agree with his usual practical nature. (One of his favorite sayings in later life was: "But does it lower the price of bread?") No, it appears to be an act of sheer romanticism, perhaps the unique romantic act of his scientific career.

Following the preface, there is an introductory section on logical notation. In it, for the first time, is found the symbol ϵ to indicate set membership and $\supset$ to indicate inclusion. Peano carefully notes the necessity to distinguish the two. The first symbol is the initial of the Greek ἐστί (is). The second is the Roman letter C turned upside-down. In making this distinction he seems to have anticipated Frege. Russell said in *Principles of Mathematics* that the distinction was known to Frege, and he gives a reference to the *Begriffsschrift*, but — as Peano himself pointed out [see Jourdain 1912, 269] — this seems to have been a mistake.

The symbol $\supset$ is also used for implication, when treating propositions. The letter C as initial of the Latin *est consequentia* (for propositions) and *continet* (for classes) is introduced solely to define its inverse $\supset$ (*deducitur* for propositions and *continetur* for classes.) He also introduces square brackets for inversion of operations. Thus $[x\epsilon]$ is 'the x's such that', a notion already considered in his *calcolo geometrico* and symbolized there by $x:$. The upside-down Greek epsilon is also introduced as a substitute for $[\epsilon]$. The symbol for 'false' or 'nothing' is Λ, which is the capital letter V (for *verum*, i. e. 'true') turned upside-down. Peano also places variable letters as subscripts on $\supset$ to indicate universal quantification.

The *Arithmetices principia* proper, which occupies 20 pages, begins with a list of four undefined terms and nine axioms. Four of these refer to the symbol =, which is here taken as an undefined term. The other five, referring to the three undefined terms: N, 1, $a + 1$, are the five universally known as

Peano's Axioms. Then, in succession, are introduced: addition, subtraction, maximum, minimum, multiplication, powers, division, some theorems from number theory, rationals, irrationals, and various theorems concerning open and closed intervals of real numbers. Altogether, a remarkable achievement.

He immediately set out to achieve a similar goal for geometry. The result, which was completed in June 1889, was not as original, but still noteworthy. But what was Peano's goal, and how did he see his earlier achievement? He says, in a note to the preface of *I principii di geometria logicamente esposti* (The principles of geometry logically presented) [18], in which he refers to *Arithmetices principia*:

The study of the principal operations of logic is due to G. Boole. Making use of the studies of Boole and of others, I succeeded as the first, in the booklet mentioned, to present a theory using purely signs having a meaning determined either by definitions or by their properties.

In the preface he states his rule for research: "Use in our propositions only terms with a completely determined value," and what was not done in the earlier booklet: "Make precise what is meant by definition and by proof."

We have in these two booklets, the modern axiomatic view, fully expressed and put into practice. This view was popularized by Hilbert's later remark: "One must be able to say at all times — instead of points, straight lines, and planes — tables, chairs, and beer mugs," but already in the 1889 geometrical pamphlet Peano was talking about "undefined" terms. The first line of his exposition is: "The sign 1 is read *point*," and in his commentary he says:

We thus have a category of entities, called points. These entities are not defined. Also, given three points, we consider a relation among them, indicated by $c \in ab$, and this relation is likewise undefined. The reader may understand by the sign 1 any category whatever of entities, and by $c \in ab$ any relation whatever among three entities of that category . . . .

The emergence of the modern view of axioms can be found in the brief period from 1882, from Pasch's *Vorlesungen über neuere Geometrie* (on which Peano based his own development) to Peano's *I principii di Geometria logicamente esposti*, from the implied position of Pasch's remark: "In fact, provided the geometry is to be truly deductive, the process of inference must be entirely independent of the figures," to the explicit stance adopted by Peano. Pasch was concerned that his axiom set be complete, that it furnish a basis for rigorous proofs of the theorems. Peano also accepted as a goal an independent set of axioms. He was, however, unable to show this for the geometrical axioms, and it was only in 1891, after he had separated the 'famous five' from the postulates dealing with the symbol = that he showed the absolute independence of his axioms for arithmetic.

In his remarks on proofs, Peano has an interesting comment on the limitations of traditional logic. "It is known," he says, "that scholastic logic is of little use in mathematical proofs, seeing that in it the classification and rules of the syllogism are never mentioned, while on the other hand, one uses arguments that are entirely convincing, but not reducible to the forms considered in traditional logic." The problem he then poses is, how to recognize a valid mathematical proof:

But this question can be given an entirely satisfactory solution. In fact, reducing the propositions, as is done here, to formulas analogous to algebraic equations and then examining the usual proofs, we discover that these consist in transformations of propositions and groups of propositions, having a high degree of analogy with the transformation of simultaneous equations. These transformations, or logical identities, of which we make constant use in our argument, can be stated and studied.

The list of the logical identities we use was made already in my [*Arithmetices principia*]; many of them were listed by Boole. Their number is large. It would be an interesting study, and it has not yet been made, to distinguish the fundamental identities, which must be accepted from the beginning, from the remainder, which are contained in the fundamental ones. This research would lead to a study of logic, analogous to that made here of geometry, and in the preceding booklet of arithmetic.

Peano is here a long way from his apparent 'combinatorial' interest, as expressed in the *Calcolo geometrico*!

This was followed by a note on Wronskians [17], in which Peano gives examples to show that the Wronskian of several functions of a single variable may be identically zero on some interval without the functions being linearly independent on that interval.

On Sunday 14 April the university began its eleven-day Easter vacation, and Peano began a hike to Genoa. He is said to have arrived on Tuesday, two days later. This is a distance of 109 miles by modern highways; Peano must have been in excellent physical condition!

Peano's next two publications were brief notes criticizing points in two new calculus texts. One gave the proof of a new form of the remainder term in Taylor's formula [19], presented without proof in the *Calcolo differenziale* of 1884. The other called attention again to the *Calcolo differenziale*, this time with regard to a theorem on Jacobians [20].

Early on the morning of 7 March, 1889, Angelo Genocchi died, two days after his 72nd birthday. Peano was asked to write the necrology for the university yearbook, and he used the occasion to tell the origin of the calculus text [21]. Peano praises his former professor and adds: "A monument worthy of him would be the collection and classification of his works, now scattered in so many different journals or not yet published." This seems not to have happened and, ironically, he is most remembered for his role in producing the calculus text that he disowned.

Also in this yearbook, the rector, Andrea Naccari, found worthy of mention that Ida Terracini (no relation to the later professor Alessandro Terracini) was enrolled in the first year of the regular course in pure mathematics. She graduated in 1892, probably the first woman graduate in mathematics at the University of Turin.

With the death of Genocchi, the way was now open for Peano's aspirations to the chair of calculus. He is listed in the yearbook as *incaricato in* (i. e., charged with) infinitesimal calculus, and a footnote explains: 'Until the professorship is awarded.' A commission was named to find a successor to Genocchi, and Peano duly presented his credentials in September. He was already in correspondence with Felice Casorati in Pavia and, believing him to be a member of the commission, he wrote on 26 October expressing his concern:

Illustrious professor,
    At first I received the announcement that the commission for the competition for the Chair of Calculus at Turin would meet in the second half of this month, and then that the competition was put off to an undetermined date. Finding myself completely in the dark here, and having a number of personal affairs to take care of, I am trusting in your well-known kindness in asking for information.
    I would like to know whether the commission will meet sooner or later, or whether a second competition will be called for and another commission appointed. Please give me any information you in fairness can, and without violating any obligations.
    Thank you for the notes you kindly sent me, and especially for the latest on the curvature of surfaces, for which I have not yet thanked you. This note interested me a great deal, especially since it turned out useful in a question I have been studying. I am taking the liberty of enclosing a summary statement of this question, along with the solution. If you believe it merits attention, I shall try to publish it in some periodical.
    Thanking you in advance for the favor, I have the honor of professing myself, your devoted

G. Peano

Casorati replied from Pavia on 4 November:

Dear Peano,
    What I know of the competition for Calculus at Turin is only what the Minister telegraphed around the middle of October, namely that the competition had to be put off to a later date because of the impossibility of finding two members to add to the commission along with Professors Arzelà, D'Arcais, and Tonelli.
    My health was then and now so bad as to force me to scrupulously obey my doctor, who has forbidden, with particular insistence, every serious intellectual work. That is

why I was unable to agree to be one of the two members mentioned above. In the meantime, no harm is being done to Turin, since the actual *Incaricato* is not lacking in ability, learning, or a gift for teaching.

I regret that my reply cannot satisfy the just desires you expressed to me.

I am pleased that my note on curvature was of interest. This is something that has happened to many other mathematicians; those few printed pages brought me a great number of very gratifying pages from my colleagues. Your formula seems to me very deserving of attention. Hence, if I were in your shoes, I would not hesitate to publish it. If you like, I or some other member could insert it in the *Rendiconti* of the *Lincei*.

As soon as I receive copies, I shall send you a reprint of my note in the *Acta* of Mittag-Leffler, with a few additional thoughts that I hope you will find opportune.

I am, with respect,

yours affectionately,
Felice Casorati

It must have been disappointing to Peano, not to have Casorati on the commission, since he was so clearly in favor of Peano, and to hear that "no harm is being done to Turin" was cool comfort! There were, after all, the "personal affairs" that he mentioned, so that the salary and security of a professorship meant much to him. Patience was required, for it would be another year before he would be named Genocchi's successor. As some consolation, he received in November a small promotion at the Military Academy to Professor of the First Class.

There was nothing to do but accept the situation, so in his next letter to Casorati, on 26 November, Peano merely thanks him for the information and moves on to another item. This other item is not another correction of a textbook mistake, but one of Peano's rare attempts to vindicate his priority to a correct formula. This was not typical of him, especially in his later life, but at this period of the competition for the calculus professorship he presumably felt that he needed all the recognition he could get. The case in point here was an approximation formula for the perimeter of an ellipse that J. Boussinesq had published in the *Comptes Rendus* of the Academy of Sciences of Paris. Peano had already given this formula, in a slightly different form, in his *Applicazioni geometriche* of 1887. He asks Casorati in his letter whether the matter is important enough and worth the effort of vindicating his priority. He also accepts Casorati's offer to submit the article, not yet completed, on the area of a surface to the *Lincei*. He says, in fact: "My delay in replying comes from my intention to send the manuscript also, but matters occupying me these days are so pressing that they absolutely prevented it. I shall send it to you as soon as possible."

One wonders what these "pressing matters" were, since a publication in the *Rendiconti* of the *Lincei* Academy would certainly favor him in the competition. Nor were his duties at the university any heavier than the year before. It must be remembered, however, that he had been doing the job of both professor *and* assistant for some time. No doubt he now felt that *he* ought to have an assistant, now that he was *incaricato*, and in fact an assistant was furnished him, but only at the beginning of the next academic year.

On 14 December, Peano sent the completed manuscript to Casorati. His anxiety about it is shown in a note written on Christmas day, only eleven days later:

Illustrious professor
On the 14th of this month I sent you my manuscript on the definition of the area of a surface; I fear it has been lost. Would you please tell me if you received it?

My remark about the approximation of the perimeter of an ellipse has already been presented to the Academy of Sciences of Paris [22].

Best wishes! Believe me

your most devoted,
G. Peano

This time Casorati answered immediately. He wrote the following day to apologize for his delay. Remarking that his health was improving, he promised to give Peano's manuscript his first attention. With regard to Peano's vindication of priority, which had been presented to the Academy of Sciences of Paris by Charles Hermite, he wrote: "I am glad your remark about the formula for the ellipse has been presented to the Academy of Paris. Boussinesq was very pleased with that formula, but he is a good man, and can only find your vindication just."

Casorati wasted no time, and could report to Peano on 6 January, 1890, that he had sent his note to Pietro Blaserna for insertion in the *Rendiconti* of the *Lincei* Academy (of which Blaserna was secretary and would be president from 1904 to 1916.) He says that he has asked that the proof-sheets be sent directly to Peano, and suggests that Peano "remind the reader of the merit of Schwarz independently of the *Cours* of Mr. Hermite."

Peano's next publication was a bombshell — the first example of a space-filling curve [24]. It is a landmark in the history of the study of dimensionality. Already, in 1878, Georg Cantor had shown that it was possible to establish a one-to-one correspondence between the points on a line and the points on a plane, thus dispelling the common notion that there are 'more' points in two-space than in one-space. E. Netto showed shortly afterwards that such a correspondence was necessarily discontinuous. It was then thought that a continuous curve in two-space, i. e. one given by continuous parametric functions of a single variable, $x = f(t)$ and $y = g(t)$, was such that it could be enclosed in a region of arbitrarily small area. In four pages of the *Mathematische Annalen* Peano gave expressions for function f, g, such that as t varies over the unit interval, the curve described goes through every point in the unit square. Thus, as Peano remarks in his first paragraph, "given an arc of a continuous curve, without any other hypothesis, it is not always possible to enclose it in an arbitrarily small region." Peano's mapping is not one-to-one, of course, since that would contradict Netto's theorem. In his *Grundzüge der Mengenlehre* (1914) Felix Hausdorff wrote of Peano's discovery: "This is one of the most remarkable facts of set theory."

In 1891 Hilbert published the first intuitive geometrical example of such a curve. His curve results as a limit of a sequence of curves. It is probable that Peano was led to the construction of his curve by such considerations. This is shown by his publication in the last edition (1908) of the *Formulario* of such a sequence of curves. He also had one of the curves in this sequence constructed on the terrace of the villa he purchased in the summer of 1891, where the curve showed up as black tiles on white. His 1890 publication, however, is purely analytic. Ugo Cassina has suggested that this is probably because he wished no doubt about the validity of his result and because he typically suppressed everything unnecessary to the goal set. "Besides," Cassina added, "the difficulty does not lie in becoming aware intuitively of the fact that a planar region can be conceived as the limit of a variable polygon, but in giving the explicit expression of the coordinates of a point of a planar region as *continuous* functions of a variable parameter in the interval."

Peano closes his note with the observation that the parametric functions are nowhere differentiable. We may add that this curve also has the property that, given any two points on the curve, the arc length between the two points is infinite. A curve often cited as having this property was invented by Helge von Koch in 1904.

Two brief notes in the Belgian journal *Mathesis* followed. The first [25] showed the conciseness of his new symbolism by presenting the 25 propositions of the Fifth Book (the theory of proportions) of Euclid's *Elements* on a couple of pages. The second [26] gave minimal conditions necessary for second order partial derivatives to commute. The theorem is:

If $f_{xy}(x, y)$ exists in a neighborhood of $x = a$, $y = b$, and is continuous at $x = a$, $y = b$, and if $f_y(x, b)$ exists in a neighborhood of $x = a$, then $f_{yx}(a, b)$ exists also, and $f_{yx}(a, b) = f_{xy}(a, b)$.

It seems that someone told Peano that Schwarz had already published this theorem, so recalling Schwarz' discovery of the error in Serret's definition of the area of a curved surface, Peano wrote him on 27 April, 1890, to ask if this were the case. Apparently Schwarz replied in the negative.

Peano seems at this time to have become convinced of the real value of symbolism in mathematics. But how to spread his ideas? No doubt journals were reluctant to accept articles requiring a great many new symbols, and it would take more than a display of Euclid in concise symbolism to make converts. To get printed *and* read an article using his symbolism would have to have important mathematical content. His next publication in the *Mathematische Annalen* did.

This paper [27] proved the existence of a solution of a system of differential equations of the sole assumption that the functions considered be continuous (a generalization of his 1886 theorem.) After stating the theorem, he says: "The whole proof has been reduced here to formulas of logic, analogous to the formulas of algebra, for, although it is not difficult, its complete development in ordinary language would be excessively complicated." He has, of course, understated his case. Only the second half of the paper is given to the proof of the theorem. The first half is given to explaining the logical symbolism and developing some theory used in the proof. The theorem was accepted by mathematicians as valid, but it is doubtful that many thought the symbolism really necessary. It is probably significant that in 1893 Gustav Mie published a popular version in German of the proof of the theorem.

It was in this work that Peano first introduced the distinction between an individual and the class composed of this individual alone. He used for this the Greek letter iota, the initial of the word ἴσος (equal to). Thus, if **b** is an individual, then ι**b** is the set containing the single individual **b**. As a consequence, the identity **a** = **b** may be written **a**ε ι **b**. This paper is of interest, too, because it contains the first explicit statement of the axiom of choice, fourteen years before its statement by Zermelo. After remarking that he had translated a certain symbol by "an arbitrary individual of the class **a**", Peano then says: "But since we may not apply an infinite number of times an *arbitrary* law whereby to a class **a** is made to correspond an individual of that class, we have formed here a *determined* law whereby to each class **a**, under certain hypotheses, we make correspond an individual of that class." Thus Peano rejects the axiom of choice. He discussed this more thoroughly in 1906 [133] (after Zermelo had introduced it in 1904), again insisting that the axiom of choice was not an ordinary form of argument and that a proof using it was not valid "according to the ordinary value of the word 'proof.'"

In July, Peano was on vacation in Savona, on the Riviera about 29 miles west of Genoa. This was followed by an attack of smallpox, which lasted from 1 to 22 August.

When the academic year began in October 1890, Peano had an assistant for the first time, Filiberto Castellano, who had previously been assistant in algebra and analytic geometry. He had graduated at the university the year after Peano and would soon become a professor of rational mechanics at the Military Academy. He early introduced vectorial methods into his lessons, which were published as a text in 1894. According to Peano's necrology of Castellano [190] (after his death on 24 January, 1919), this text was the first in rational mechanics to make methodical use of vectors.

CHAPTER 5

EXTRAORDINARY PROFESSOR

On 1 December, 1890, Peano's long wait was over. After regular competition, he was named Extraordinary Professor of Infinitesimal Calculus at the University of Turin. With the appointment secured, he could now move more confidently with the plans he had been maturing for founding his own journal. It had been a decade since his graduation, a decade of struggle in which he was given the tasks of a professor, without the recognition. Apart from the brief unpleasantness over the *Calcolo* of 1884, there had been no real difficulties, and surely no one was surprised at his nomination as professor. It had, in fact, been a very productive decade, but now a new period was beginning and its first decade would be even more productive.

Before the year ended, two more notes of Peano were published. On 7 December, 1890, Eugenio Beltrami presented to the *Lincei* Academy a note on approximations to the area of an ellipsoid [28] that gave several improvements on approximations contained in a text of Boussinesq earlier that year. The second note [29] also was concerned with correcting errors, this time in von Staudt's *Geometrie der Lage*, an Italian translation of which had been published the previous year by Mario Pieri. Peano's note was presented to the Academy of Sciences of Turin on 28 December by Corrado Segre. Although Segre was five years younger than Peano, he had graduated at the University of Turin in 1883 while not yet twenty years old, and was able to win the competition for the Chair of Higher Geometry in 1888, becoming a member of the Academy of Sciences a year later. This was the last paper by Peano that would be presented to the Academy by someone else, for on 25 January, 1891, Peano was elected a member. (The election was confirmed by a royal decree of 5 February and announced at the session of 22 March.)

Peano's continued interest in the geometrical calculus showed in the publication that spring (the preface is dated: January 1891) of a 42-page pamphlet on the elements of geometrical calculus [30]. A German translation by Adolf Schepp was published later that year in Leipzig.

Plans were now ripe for the inauguration of the journal *Rivista di Matematica*. There were several 'founders' and many collaborators, but the inspiration and much of the practical editorial work was due to Peano, who was officially the Director. In this journal (and its offshoot, the *Formulario*) were published much of Peano's pioneering work in mathematical logic. The first volume (1891) contains, by Peano, five original articles, observations on another article, a reply to a 'declaration', an open letter, and four book reviews.

The founders of the *Rivista*, those who helped finance the first volume, besides Peano, included Filiberto Castellano, already mentioned, Francesco Porta, and Francesco Porro. Francesco Porta was himself a mathematics graduate of the University of Turin and had been teaching at the Military Academy since 1877. Francesco Porro, after graduating in physics at the University of Pavia in 1882, became an assistant at the Brera astronomical observatory in Milan. In 1887 he transferred to the observatory of the University of Turin, being named director on the death of its former director, Alessandro Dorna, in August of that year. Shortly after, he also assumed Dorna's astronomy course at the university (which had been temporarily entrusted to N. Jadanza), being named Extraordinary Professor in 1896. In 1901 he transferred to Genoa.

The first article in the *Rivista di Matematica* is a ten-page summary of Peano's work to date in mathematical logic [31]. In his footnotes to the article Peano calls attention to Leibniz' prediction of a language that "will be universal, equally easy and common, and able to be read without any sort of dictionary, and at the same time a basic knowledge of all things will be assimilated." He then describes some of the work of others and, especially, his own, and he says: "It thus results that the question proposed by Leibniz has been completely, if not yet perfectly, resolved." The statement is a bit strong and contains much wishful thinking, but it is revealing of Peano, for it gave him the greatest pleasure to see his work as a fulfillment of the dream of Leibniz. Here he also quotes Ernst Schröder to back up his claim. For the first time he mentions Gottlob Frege in a footnote.

A brief note summarizes Books Seven, Eight, and Nine of Euclid's *Elements*, in a way similar to what he had previously done for Book Five [32].

An article entitled simply 'Formulas of mathematical logic' [35] is his first attempt to treat logic along the lines previously attempted for arithmetic and geometry. It is an analysis of the calculus of propositions based on four undefined or *primitive terms* (represented by the symbols $\supset \cup - \Lambda$) and twelve axioms or *primitive propositions*. This was followed later by brief 'additions and corrections', in which he was able to reduce the number of primitive propositions and add a number of new formulas, giving credit for most of them to Volumes I and II of Ernst Schröder's *Vorlesungen über die*

Algebra der Logik, which appeared in 1890 and 1891 respectively. In the meantime, he wrote a review of these two volumes [39]; althought he has criticisms to offer, it is on the whole full of praise. Peano says: "I conclude this review with admiration for this magisterial work of the author, in which is contained everything that refers to this new branch of science up to today."

The important article 'On the concept of number' [37] is likewise in two parts, the first following the 'Formulas of mathematical logic.' In it he compares his analysis with Dedekind's as given in *Was sind und was sollen die Zahlen?* (1888) and proves, for the first time, the independence of his five axioms for the natural numbers (the positive integers.) The axioms, as given in [37], are:

(1) $1 \in N$.
(2) $+ \in N \backslash N$.
(3) $a, b \in N . a+ = b+ : \supset . a = b$.
(4) $1 - \in N+$.
(5) $s \in K . 1 \in s . s+ \supset s : \supset . N \supset s$.

They may be read:

(1) One is a number.
(2) The sign + placed after a number produces a number.
(3) If **a** and **b** are two numbers, and if their sucessors are equal, then they are also equal.
(4) One is not the sucessor of any number.
(5) If s is a class containing one, and if the class made up of the sucessors of s is contained in s, then every number is contained in the class s.

These postulates were not further reduced in number, although their form was changed, due to modifications in Peano's notation.

With regard to the possibility of defining *number*, he says:

It is necessary to first say which are the ideas we may use. Here we suppose known only the ideas represented by the signs $\cup$ (and), $\cap$ (or), $-$ (not), ϵ (is), etc., which were treated in the preceding note. Now, *number cannot be defined*, since it is clear that however these words are combined, we will never have an expression equivalent to *number*.

He takes up this question again in the second part of this article, where he considers the questions: "Can the idea of number be defined, using simpler

ideas? Can the commutative property be deduced from simpler properties?"
His reply is:

> To these questions may be given different answers by various authors, since simplicity
> can be diversely understood. For my part, the answer to the first is that number (posi-
> tive integer) cannot be defined (seeing that the ideas of order, succession, aggregate,
> etc., are just as complex as that of number.) The answer to the second has been affir-
> mative.

It is in the concluding paragraphs of this article that we find the first hint of
the *Formulario* project that would take up much of his editorial activity
from 1894 to 1908. This will be discussed in Chapter 6.

The book reviews reveal Peano's linguistic ability, for the books reviewed
were written in English, Italian, German, and Polish. The first [33] reviews
The directional Calculus, based upon the methods of H. Grassmann (1890)
of Edward Wyllys Hyde. This was the first textbook on Grassmann's calculus
in English. Peano finds it similar to his own *Calcolo geometrico* of 1888 and
with some of the same defects of terminology. The second [34] reviews a
calculus text of Francesco D'Arcais. Peano finds it a good combination of
rigor and clarity. In fact, this book reached a third edition in 1912. The third
[38] is not properly a review. Peano begins: "Because of the importance
and topicality of the subject, we announce the appearance of this book,
publishing a summary and postponing the review of it until after having
carefully read it." The subject of this volume is 'The theory of operations'
and its author was Samuel Dickstein, with whom Peano would later share
an interest in an international auxiliary language. Peano closes: "The difficulty of the language (Polish) is more apparent than real, since formulas,
mathematical terms are already common to all languages, and the style
peculiar to mathematicians greatly help in reading it." The fourth review
[39], of Schröder's *Algebra der Logik*, has already been mentioned.

Along with an article by C. Segre on directions of geometrical research,
Peano published a few observations [36], that were suggestions made to
Segre after he had submitted his manuscript in February. Segre did not
revise his manuscript and Peano simply appended his remarks. This gave
rise to a brief polemic, Segre writing a 'declaration' of his position and
Peano appending another 'reply' [38.1]. Peano had pointed out that some
of Segre's theorems were incorrect in that they had exceptions which he
had not pointed out. Segre defended himself by saying that the moment of
discovery is more important than the rigorous formulation, and mentioning
works that were historically important, but contained errors.

Peano took personally one of Segre's remarks and replied: "I hold the scientific polemic to be one of the forms under which ideas may sometimes usefully be expressed. I, myself, owe to polemics a close friendship with a momentary adversary whom I did not previously know. But, in order that the polemic not be *useless and damaging*, it is necessary that it be carried out strictly scientifically." The friendship referred to is probably that with Francesco Giudice, following their brief exchange in 1888. Giudice became a collaborator on the *Rivista di Matematica* and Peano presented several of his papers to the Academy of Sciences of Turin. In Segre's case, this may have been the wedge that began the split of the mathematics faculty at the university into two factions, with Peano the leader of one side and Segre a leader of the others.

Peano (as Director of the *Rivista*) had the last word: "I believe it new in the history of mathematics that authors knowingly use in their research propositions for which exceptions are known, or for which they have no proof; and instead of following past mathematicians in the innumerable points in which they worked correctly, take as model the rare points that are defective."

The open letter to Giuseppe Veronese (dated 1 December, 1891) [40] is in response to a criticism in Veronese's book *Fondamenti di geometria a più dimensioni*. There is also brief mention of the polemic with Segre, which had prompted Veronese's remarks, and Peano challenges Veronese to answer publicly. This Veronese does in a note dated 29 February, 1892 (but with a footnote saying that an answer would have been published sooner except for a serious family misfortune.) The lengthy reply is a strong defence of his book. Peano's brief reply [54] (dated 22 May 1892) asserts only that Veronese did not directly answer his objections. He closes with: "The purpose of my open letter was to appeal to the public against the pronouncements made by Prof. Veronese with regard to my writings. My purpose was obtained, the question is closed."

The last of Peano's articles in the first volume of the *Rivista* is merely a footnote to a proof by Emmanuel Carvallo of the "fundamental theorem of spherical trigonometry" [41]. Peano points out that this is an elegant application of the internal product of two vectors and that Carvallo follows Grassmann's notation $A \mid B$ for this product.

At the 22 November session of the Academy of Sciences of Turin, Peano presented his first note since becoming a member [42]. In it he continues the discussion of Taylor's formula begun in 1889 [19], considering it for the case where the function involved is expandable by a Taylor's series. A

German translation of this note was appended to the 1899 German translation of the *Calcolo differenziale* of 1884.

There was one more publication in 1891, a formula dealing with the area of an ellipsoid, that was proposed as a problem for readers of the *Nouvelles Annales de Mathématiques* [42.1].

On 14 July, 1891, Peano bought a villa in Cavoretto, a district in the hills just south of Turin and east of the Po River (specifically at Val Patonera, No. 566.) This included 8820 square meters of land (a bit over two acres) and cost 7500 lire.

As a result of disorders, the university was closed from 26 January to 3 February 1892. The disorders were then renewed and several courses had to be suspended also until 15 February. Since this interfered with the examination period, the Academic Council granted an extraordinary session of exams for those students who needed to pass only two more in order to graduate.

No doubt Peano took advantage of the unexpected university vacation to continue his intensive activity on the *Rivista di Matematica*. Volume 2 (1892) contains no less than eleven items written by him. These include an observation on a problem submitted to the *Rivista* [43], a summary of Book Ten of Euclid's *Elements* [44], interpreting the propositions algebraically and using Peano's symbolism, a review, written with Francesco Giudice, of an elementary algebra text of Domenico Amanzio [44.1], observations on the analysis text (in seven volumes, 1885–91) of Hermann Laurent, professor at the Ecole Polytechnique in Paris [46], and a necrology of Enrico Novarese [46.1], who died prematurely on 14 January, 1892. (He was born on 15 June, 1858, and so was almost exactly Peano's age.) Novarese was professor of rational mechanics at the Military Academy and assistant at the university. He published a dozen articles (seven of them in the *Atti* of the Academy of Sciences of Turin) and was one of the founders (according to Peano) of the *Rivista*, to which he had contributed one article and a necrology of Sofya Kovalevskaya.

Further articles by Peano in this volume of the *Rivista* include: simple examples of increasing functions that are discontinuous in every interval [47], a proposed question [48], an elaboration of Cantor's proof of the impossibility of segments of constant, but infinitesimal, length [51], a note on the definition of the limit of a function [52], and a review of a pamphlet by Albino Nagy [53].

In the article on the limit of a function he proposed the adoption of what has since come to be called in topology 'cluster point' of a function, in place of the usual definition of limit. In modern notation, if f is defined on an open interval containing a, then 'L is the limit as x approaches a' means:

$$\forall \epsilon \; \exists \delta \; \forall x \; (0 < |x - a| < \delta \Rightarrow |f(x) - L| < \epsilon).$$

On the other hand, 'C is a cluster point of f as x approaches a' means:

$$\forall \epsilon \; \forall \delta \; \exists x \; (0 < |x - a| < \delta \Rightarrow |f(x) - C| < \epsilon).$$

Whereas there can be only one limit point, there may be more than one cluster point. This disadvantage is more than made up for, according to Peano, by the advantage that, if $+ \infty$ and $- \infty$ are allowed as values, then f always has a cluster point. Then, if the function has a unique cluster point, it is the limit point according to the usual definition. He developed the theory of limits in this way in his calculus text for the Military Academy, published in 1893 [60] and in an article for the *American Journal of Mathematics* in 1894 [68].

In an article earlier in the spring, Peano had discussed the relation between two different definitions of derivative [45]. This was a follow-up to his polemic with Gilbert, of the University of Lorvain, and was appropriately published in a Belgian journal (*Mathesis*). A note in the French journal *Nouvelles Annales de Mathématiques* [49] improves on a previous note of Emile Picard regarding the existence of solutions to systems of differential equations.

On 27 March, 1892, Peano presented his second paper [50] to the Academy of Sciences of Turin. In it he generalizes Simpson's formula to an infinite sequence of formulas, of which the trapezoidal rule and Simpson's rule are the first two. He repeats his expression for the remainder terms in these two formulas, that he had given in his *Applicazioni geometriche* of 1887, and also gives the remainder terms for the next two formulas in the sequence. Ignoring the Newton-Cotes formulas, he believes he is the first to give this type of generalization. At the session of 1 May he presented a note of Francesco Giudice, their differences apparently being forgotten. This was the first of many notes of others that he would present – and several were by Giudice.

The five students receiving degrees in mathematics from the university this year included Ida Terracini, whose enrollment was earlier mentioned (she was probably the first woman to receive a mathematics degree), and Gino Fano, who was the only one to receive 'highest honors'. Fano stayed one more year at the university, as assistant to D'Ovidio. He returned to Turin in 1901 as professor of projective and descriptive geometry, remaining there until 1938, when he was forced to leave because of the anti-Jewish laws.

In July, Peano is again the watch-dog, pointing out, in a letter to the editor of the *Nouvelles Annales de Mathématiques* [55], that a proof in an article by Hermann Laurent was not correct.

In the August issue of the *Rivista*, Peano returns to the attack on Veronese with a sharply critical review of the book of Veronese that prompted the original polemic [56]. He begins by criticizing the author's grammar and closes with: "We could continue at length enumerating the absurdities that the author has piled up. But these errors, the lack of precision and rigor throughout the book take all value away from it." That was strong enough to give rise to histrionics to match Veronese's mother's cousin, the actress Eleonora Duse!

In the fall, Giovanni Vailati replaced Castellano as Peano's assistant. He would soon become one of the most active contributors to the *Formulario* project. The opening of the academic year was delayed, however, because of the general elections. The examination period, with which the year should have begun on 15 October, was postponed to 16 November, following a telegram of the Minister of Public Instruction sent out on 13 October. In December, Peano was made Professor of the Third Class at the Military Academy.

In 1893, Giacomo Puccini, who was almost exactly Peano's age, had his first real success with 'Manon Lescaut', which had its premiere at the Royal Theater of Turin on 1 February. Peano and his wife probably attended one of the performances, since Carola was fond of music and they often went to concerts. Peano, however, did not entirely share her tastes and is known to have fallen asleep in the theater.

In the February issue of the *Rivista*, Peano follows up a letter of Giuseppe Bardelli with a calculation of the remainder term for a quadrature formula proposed by Bardelli [58].

On 22 April, Peano bought another 2065 square meters of land (about one-half acre), paying 400 lire for it.

The fall saw the arrival in Turin of Vito Volterra, who had won the competition for the Chair of Rational Mechanics at the University of Pisa when only twenty-three years old, ten years earlier. When Francesco Siacci, who had been named a senator, asked to be transferred to Naples so that he could more easily fill his office as senator in Rome, the mathematics faculty of the University of Turin on 11 July, 1893, decided to ask the Minister of

Public Instruction to invite Volterra to fill his place as Ordinary Professor of Rational Mechanics. (Siacci was then named Honorary Professor of the University of Turin.) The controversy Peano had with Volterra will be discussed in Chapter 8.

Peano's calculus lessons at the Military Academy were published in the fall, in two volumes [60]. (Parts had been printed earlier. Peano wrote to H.A. Schwarz on 4 April, 1893, apologizing that the set sent him was incomplete.) This work contains few new results, but is notable for its clarity of presentation, presumably reflecting his lectures in the classroom. It marks the end of a period in his classroom technique and suggests the beginning of another. The first five chapters (Volume 1) cover the calculus of functions of a single variable: derivatives, integrals, and series. The language used is traditional, with little of the symbolism that he had already begun using in his research. Chapters Six through Nine (Volume 2) treat vectors, functions of several variables, and differential equations. The treatment of vectors is in the Grassmann tradition, and somewhat wider use is made of his new symbolism. Parts of Chapter Six, dealing with 'number complexes', was translated into German and appeared as an appendix to the German edition in 1899 of the *Calcolo* of 1884.

Peano's reputation as a teacher seems to have reached across the Atlantic. On 4 November, 1893, he wrote David Eugene Smith at the Michigan State Normal School in Ypsilanti, Michigan, apparently in reply to a letter from Smith. In comparing his students at the Military Academy with those at the university, it is interesting that he finds those at the Military Academy much better prepared than the university students. This he attributes to the fact that the Military Academy has an entrance exam in mathematics, whereas the university admits anyone who completes the *liceo*. A few are good, he says, but in general they lack a knowledge of elementary mathematics, especially trigonometry.

Besides the note [58], the text [60], and a referee's report [57], Peano's other publications of 1893 were three brief reviews, all in the *Rivista di Mathematica*: of a note of Nicodemo Jadanza, professor of geodesy at the University of Turin [58.1], a note of Emmanuel Carvallo [59], and a textbook of Alexander Ziwet, *An Elementary Treatise on Theoretical Mechanics. Part 1, Kinematics* (New York, 1893) [61]. Peano praises this book for its form and content, and also for "the superb manner in which it is printed, something that we, unfortunately, are little accustomed to."

The small number of articles this year is explained by work on the Formulario project. We must now go back and pick up this story.

CHAPTER 6

THE *FORMULARIO* PROJECT

The first hint of the vast *Formulario* project came in 1891. In the concluding paragraph of his 'Concept of Number' [37], which contains several equations written symbolically, Peano says:

It would also be very useful to collect all the known propositions referring to certain parts of mathematics, and to publish these collections. Limiting ourselves to arithmetic, I do not believe there would be any difficulty in expressing them in logical symbols. Then, besides acquiring precision, they would also be concise, so much so, probably, that the propositions referring to certain subjects in mathematics could be contained in a number of pages not greater than that required for the bibliography.

The transformation into symbols of propositions and proofs expressed in the ordinary form is often an easy thing. It is a very easy thing when treating propositions of the more accurate authors, who have already analysed their ideas. It is enough to substitute, in the works of these authors, for the words of ordinary language, their equivalent symbols. Other authors present greater difficulty. For them one must completely analyse their ideas and then translate into symbols. Not rarely it is the case that a pompously stated proposition is only a logical identity or a preceding proposition, or a form without substance.

The *Rivista di Matematica* will try in the coming year to publish collections of this type. Hence, we invite readers to write them and to send them to us.

There seems to have been some response. At least those most closely connected with him in Turin were interested, such as Castellano and Vailati, so that in March of the following year lists of formulas were printed as a supplement to the *Rivista*, to which the following statement was added (and later repeated in the *Rivista*):

Of the greatest usefulness would be the publication of collections of all the theorems now known that refer to given branches of the mathematical sciences, so that the scholar would have only to look into the collection in order to find out what had been done concerning a certain point, and whether his own research was new or not.

Such a collection, which would be long and difficult in ordinary language, is made noticeably easier by using the notation of mathematical logic; and the collection of the theorems on a given subject becomes perhaps less lengthy than its bibliography.

The tables that follow contain a first attempt at this collection. Part 1 contains theorems, identities, etc., that refer to operations on real numbers. Part 2 contains the elements of the theory of numbers.

The notations used are summarized here. Their precise meaning has been explained in several articles in Volume 1 of the *Rivista*.

The tables are, for now, only printing proofs. The numeration of the propositions and the division into sections are provisory.

We are making a warm appeal to the readers of the *Rivista* to examine these formulas and to communicate to the *Rivista* all those that they know and do not find contained in the list, as well as corrections and observations, as the case may be.

Later these formulas will be printed separately. The subscribers to the *Rivista* will receive them free.

We shall be most grateful to readers who wish to help us in this work, by collecting propositions (with or without proofs) of other parts of mathematics.

To avoid confusion, the word *formulario* should be explained; but this is not easy, since Peano used it ambiguously. The Italian word *formulario* translates 'formulary', i. e. a collection of formulas, but is used more frequently in Italian, although it is not a common word. Peano used it in this sense, as well as for individual titles of publications in an ongoing project that was also called *Formulario*! Nor are the five editions of the *Formulario* editions in the usual sense of the word. Each is essentially a new elaboration, although much material is repeated. Moreover the title and language varied: the first three, titled *Formulaire de Mathématiques*, and the fourth, titled, *Formulaire Mathématique*, were written in French, while *Latino sine flexione*, Peano's own invention, was used for the fifth edition, titled *Formulario Mathematico*. To these must be added the *Notations de logique mathématique* (*Introduction au Formulaire de Mathématiques*) [66] of 1894. Indeed, Ugo Cassina lists no less than twenty separately published items as being parts of the 'complete' *Formulario*! These include, besides the items mentioned here, those publications in which Peano develops mathematical logic, beginning with the preliminary chapter to the *Calcolo geometrico* of 1888. Thus, the development and use of mathematical logic is the guiding motif of the project.

This project, for all the enthusiasm it aroused in a certain group of readers of the *Rivista*, was not universally popular. Most important, and no doubt to Peano's disappointment, there was very little interest in it among the professors at the university, and Peano's apostolic zeal in promoting it, in the classroom and out, soon began to alienate him form the rest of the faculty. Support for the project came mostly from recent graduates, some of whom were his assistants at the university: F. Castellano, G. Vailati, and C. Burali-Forti. Castellano and Burali-Forti were also colleagues at the Military Academy. Mario Pieri, also at the Military Academy, gave moral support, but was not a major contributor to the first edition [71], since it did not contain a chapter on geometry. This edition is often referred to as

'F 1' or 'F 1895'. Not contained in this edition, as one might have expected, was the so-called *Introduction*, a booklet of 52 pages dealing with the notations of mathematical logic, that Peano published in 1894 [66].

The *Introduction* opens with the name 'Leibniz' which would become a commonplace with Peano. Leibniz may or may not have been Peano's original inspiration − probably not − but more and more he became the inspiration for the project. Peano delighted in seeing his work as a continuation of that of Leibniz, and he would soon see it as a fulfillment of Leibniz' dream. "Leibniz," he writes, "stated two centuries ago the project of creating a universal writing in which all composite ideas would be expressed by means of conventional signs for simple ideas, according to fixed rules." After quoting Leibniz, he then moves to the nineteenth century and lists a number of people who helped in the development of mathematical logic: Boole, Cayley, Clifford, Delboeuf, De Morgan, Ellis, Frege, Grassmann, Günther, Halsted, Jevons, Liard, Macfarlane, McColl, Nagy, Peirce, Poretskii, Venn, and Schröder, to whose *Algebra der Logik* Peano refers for a more complete list.

He then mentions the *Formulario* project: "A Society of Mathematicians is now publishing a formulary intended to contain all the known propositions on certain subjects of mathematics. This formulary, written entirely in symbols, is being published by the *Rivista di Matematica*. There have already appeared the formulas of logic, elementary algebra, arithmetic, measure theory, and point set theory; and the theory of limits, of series, of continuous functions, of derivatives, etc., is being printed." Thus, a great deal of progress had already been made, although some of this was provisional; the official first volume of the *Formulario* would not appear until the beginning of 1895. Various parts had already been distributed in extract, however, the part dealing with mathematical logic as early as January 1893, and Peano apparently felt they were widely enough distributed to give references to them. There was even an Italian version of the theory of point sets, written in collaboration with G. Vivanti, that was published as a separate pamphlet before the end of 1894 [70.3].

Some of the ideas and notations of the *Introduction* were not included in the *Formulario*, but much of it was already standard, such as the symbols: $\epsilon, \supset, =, \cap, \cup, -, \Lambda$. We get an idea of Peano's linguistic ability in his discussion of functions, where illustrations are given from Greek, English, German, and even Chinese. Further, a footnote referring to an article of Poretskii would seen to indicate some knowledge of Russian. The name 'definition by abstraction' also seems to have been introduced for the first time in this booklet.

The *Introduction* was reviewed the following year in the *Zeitschrift für Mathematik und Physik* by M. Cantor. After remarking on the growth of mathematical knowledge, such that one man can no longer know it all, he mentions that the *Rivista* is publishing a *Formulaire de Mathématiques* to collect together the formulas of mathematics and that this can be done by using the symbols of mathematical logic. He then says:

> To be sure, there is a nuisance here, namely that not every mathematician is able to read this writing! But, for good or bad, one will have to learn to understand these symbols, which are supposed to realize Leibniz' general writing. Writers are beginning, at least in Italy, and apart from the *Rivista di Matematica*, to use it, such as, for example, Mr. Francesco Giudice in the Acts of the Turin Academy of 31 December, 1893. Of course one must have some help in learning a language. A grammar, a dictionary must be at hand, and as such will serve this small book, of whose publication we are informing our readers. That Mr. Peano, more than any other, is qualified to publish such an introduction to a knowledge of the symbols of mathematical logic requires no argument, and a glance into the writing itself will convince anyone that the right man was chosen for it.

Volume 1 of the *Formulaire de Mathématiques* [71] was the result of this first round of activity, the culmination of over two years of work on the part of Peano and several others. The text consists of nine chapters, whose titles and authors are: 1. Mathematical logic (G. Peano and G. Vailati) [72]; 2. Algebraic operations (G. Peano and F. Castellano) [73]; 3. Arithmetic (G. Peano and C. Burali-Forti) [74]; 4. Measure theory (C. Burali-Forti); 5. Classes of numbers (G. Peano) [75]; 6. Set theory (G. Vivanti); 7. Contributions to the theory of algebraic numbers (G. Fano).

Peano must have felt strongly the imperfection of this volume, for he wrote in the preface:

> But, whatever the care taken by the author of some part, one can find there mistakes and omissions, especially in a first edition. On the other hand, it is almost impossible for one person to examine all books, to verify all the historical indications. We shall publish in the *Rivista di Matematica* and in the succeeding volumes of the *Formulaire* all the additions and corrections that are pointed out to us. We can even republish a part that has already appeared, after it has been perfected.

To assist the future collaborators, he gives a long list of rules to be followed. Some of them are worth quoting:

(1) The single law that governs the notations of the *Formulaire* is that they must be the simplest and most precise to represent the propositions with which they are concerned.

(2) The notations are somewhat arbitrary, but the propositions are absolute truths, independent of the notations adopted.

(3) Every time a new theory is translated into symbols, new signs will be introduced to indicate the new ideas, or the new combinations of preceding ideas, that are met in this theory.

(4) The reduction of a new theory into symbols requires a profound analysis of the ideas that occur in this branch. Imprecise ideas cannot be represented by symbols.

(5) A new sign should not be used to represent each idea indicated by a simple word in ordinary language. This point constitutes an important difference between ordinary language and the notations of mathematical logic.

(6) A new notation will be introduced by means of a *definition* when it brings a notable simplification. A new notation will not be formed when the same ideas can already be simply represented by the preceding notations. ...

(7) A new notation will be introduced only if the simplicfication that it brings will be used in the propositions following. Definitions alone do not make a theory.

(8) It is well to form a table of the words of ordinary language that cannot be translated by a simple symbol, but that must be decomposed into a combination of signs.

(9) If care is not taken, the same propositions will be written several times, under different forms whose identity is not immediately recognized. Sometimes a proposition is presented as a theorem, which is only an identity.

(10) Every definition is expressed by an equality; in the first member is a sign to be defined, which is a new sign or a new combination of known signs; in the second is its value. It is necessary that the two members be *homogeneous*; they must contain the same variable letters. Nevertheless, variable letters can be left unexpressed if they have the same constant value in the propositions following.

. . . .

(12) There is no possibility of confusion in representing by the same letter different ideas in different propositions, for at the beginning of each proposition it is necessary to give the meaning of the variable letters.

. . . .

(19) There are six signs of logic that occur between classes or between propositions, $\cap$, $\cup$, $-$, $=$, $\supset$, Λ; in Part 1 of the *Formulaire* the primitive ideas are $\cap$, $-$, $\supset$, and the others are defined. Other combinations are possible. In the manuscript, it is best to give the sign for 'not' the form $\sim$, so as not to confuse it with $-$ (minus).

(20) After having written a formula in symbols, it is useful to apply several logical transformations to it. It can thus be seen if it is possible to

reduce it to a simpler form, and one can easily recognize if the formula has not been well written.

(21) This is because the notations of logic are not just a shorthand way of writing mathematical propositions; they are a powerful tool for analysing propositions and theories.

. . . .

(23) In enumerating the propositions of a section, it is good not to use a continuous sequence of numbers, but to leave some gaps where propositions are later to be inserted.

(24) It is good to copy the historical indications that are found in books without verifying all of them, for this requires a lot of work and can be done by another collaborator.

(25) The proofs of the propositions, or at least the connecting links between propositions, can also be published, but the transformation of a proof into symbols is in general more difficult than the statement of the theorem.

The collaboration of Vailati, Castellano, and Burali-Forti on the first three chapters (presented, in fact, without the authors being named) especially regard the historical notes. Apropos this and rule No. 24 above, we may present here what Peano wrote two years later, when he wished to assume responsibility for the historical notes also:

Historical indications, whether for propositions or notations, are always useful, and especially so in the *Formulario*, because they give the reader a rest and bring out better the importance of the propositions and often the advantage of the ideography. But they require much work in order to have any value. The indications that are found in the books of the preceding generations, and which are still found in some modern books, according to which, for example, algebra is due to Descartes or Viète or to the Italian mathematicians of the 13th century or to the Arabs or to the Indians or to the Greeks, have no precision. Nor does, for example, the more precise statement that the use of letters is due to Viète, since they are found in Aristotle and were commonly used by Euclid, as the citations in the *Formulario* show.

Nor can those indications be useful for us which are of the form: "The sign + is due to Leonardo da Vinci." The examination of such a statement leads to a lot of work, and what results from it is that this sign was in fact used by this scientist to indicate the numeral 4. To attribute to him the use of the sign + is a good as attributing it to the first caveman who carved a cross. Citing Widmann is similar to this.

Then, too, the indications that give the precise name of the author and the page of the book cited are often not exact because of the errors accumulated by passing through many hands, and we may look in vain in the place indicated for the passage in question. At other times we find it, but with a completely different meaning from that given it by the historian.

In consequence one must go back to the original of the passages cited. The citations in the *Formulario* give precise indications, so that anyone can easily look it up in the book cited, and often the very passage is given. This is done as much as possible, because even in the *Formulario* some citations need to be made more precise.

It should be noted that the historical indications contained in the *Formulario* do not at all pretend to go back to the first origins of the proposition in question, but only to indicate an author in whose work it may be found. A further study may always substitute for it other citations relative to an earlier epoch.

After the publication of the first edition of the *Fomulaire,* work was immediately begun on a second edition. This would appear in three parts published separately in 1897, 1898, and 1899. But Peano's activity was not confined to the *Formulario* project, so we now return to other facets of his work.

CHAPTER 7

ORDINARY PROFESSOR

In January 1894, the university was again disrupted by a typical student disturbance. According to the rector, the trouble started with a false notion that the Academic Council had vetoed a proposal for a special examination period, permission for which, at any rate, would have had to come from the Minister of Public Instruction. So, on Monday, 22 January, and the following day the students "began to shout, stopped classes, forced their way into the Great Hall, breaking down the door, and shortly after left, making a great noise, and began their usual round of the city with the purpose of halting classes in the various institutes connected with the university." As a result, the university was closed on 23 January and reopened on 8 February, following the publication on 5 February of a manifesto of the rector advising the students that he had government authorization, should the university have to be closed again because of disturbances, to call off the summer examination session as well as take other steps thought necessary. It apparently worked; there were no more disturbances that year, and on 4 May the Academic Council set dates for both the summer and fall examination periods.

Partly because of work on the *Formulario*, Peano's publications of 1894 were mostly minor. After showing how Taylor's theorem applies to linear systems [62] and commenting on the point set theory of the *Formulario* [63], he published a lengthy study of the foundations of geometry [64]. Although this last contains some new items, e. g. the extension of the geometry of position to metric geometry using 'motion' as an undefined term, it is principally an elaboration of the 1889 booklet *I principii di geometria logicamente esposti* [18]. Apparently answering criticism, he admits that his work of 1889 was based on Pasch, but points out that there were differences. He then says that the first scientific question presented is whether the postulates are independent, but while he gives several proofs that some are independent of others, he admits that he is unable to prove the independence of his postulates. The work also contains a curious echo of the polemic with Veronese, in an observation on the "practical, or experimental, nature of the postulates." He writes:

In order that this work merit the name of geometry, it is necessary that these hypotheses

51

or postulates express the result of the simplest and most elementary observation of physical figures. Geometry of position, or projective geometry, then, is a part of general geometry; hence its postulates must be found among those assumed for general geometry.

Consequently, from the practical point of view, it does not seem to me to be licit to assume, for example, the following as a postulate on which to base projective geometry: 'Two lines in the same plane always have a point in common,' because this proposition is not verified by observation; rather, it is in contradiction to the theorems of Euclid.

Projective geometry starts out from the postulates of elementary geometry and, with opportune *definitions*, introduces new entities, called ideal points (both in Euclidean and nonEuclidean geometry), with the result that the new entities obtained satisfy all the preceding propositions.

The death of E. C. Catalan, of the University of Liège, on 14 February, was the occasion for publishing a letter from him, that Peano had earlier not meant to publish [64.1]. (It was dated 5 February, 1892.) Peano indicates that in January Catalan had sent corrections and additions to the *Formulario*. The published letter contains the very nice compliment: "I would like to be able to *understand* Italian as well as you *write* French."

Also in a brief note in the *Rivista* [65], Peano points out that Ludovico Richeri was a precursor of mathematical logic, and quotes from a publication of 1761.

In August 1894, Peano attended the 23rd meeting of the French Association for the Advancement of Science, held in Caen (Normandy), and on Saturday 11 August he presented a paper on 'Notions of mathematical logic' [67]. This was the first of several papers he would present in foreign countries, from Switzerland to Canada. The paper read at the Caen conference is a typical propaganda piece, designed to persuade others to use the symbolism of mathematical logic developed by him, as is also the note 'On the definition of the limit of a function. An exericse in mathematical logic' [68] published in the *American Journal of Mathematics*. Worth noting in the Caen paper is the suggestion that other journals should publish lists of the theorems of various branches of mathematics in the way the *Rivista* was doing. The American paper concludes with a remark that shows us how Peano considered mathematical logic: "The two objects of mathematical logic, the formation of a symbolic writing and the study of the forms of transformations, or of reasoning, are strictly connected."

At the Caen conference the 'order of the day' for Sections 1 and 2 on 14 August was "the study of the means that will be of a nature to assure an easier and more frequent exchange of views among mathematicians of various nations. and which will be able to thus contribute to the progress of the mathematical sciences and to the perfection of its methods." A great number of people took part in the discussion, following which nine resolutions were passed unanimously.Two of them are of special interest here. The first says that full support is given to the idea of creating an international mathematical congress. Resolution No. 5 says: "It is considered that the great efforts made by Professor Peano and several of his colleagues for the propagation of *mathematical logic* and the publication of a *mathematical formulary* are of a nature to strongly contribute to the goal for which they are intended."

Peano returned to Turin to find a letter from Felix Klein waiting for him. It would appear that Klein thought fascicle 10 of the *Rivista di Matematica* had not been sent to him, for Peano replied on 29 August, 1894, that No. 10 was missing only through a typographical error. He then discusses the *Formulario* project in an apparent attempt to interest Klein — and others — in it. He remarks:

I have set myself the task of publishing the *Formulario* and I am happy to have the collaboration of a few colleagues and of several young recent graduates, who have enthusiastically taken over the various parts. But my efforts are intended to make these methods known to the scientific world. Just now at the Caen congress there was a vote of praise for the *Formulario*, but I would desire others to occupy themselves with this very important subject. ... Hence, I shall not cease to work with all means until the importance of this question be sufficiently recognized.

This seems to have interested Klein sufficiently for him to enquire further, for Peano wrote again on 19 September, in reply to another letter from Klein. He describes the procedure used in publishing the *Formulario* and offers to send copies of the 'Introduction' [66] if Klein will distribute them at an upcoming congress in Vienna.

Peano also extended his propaganda campaign to the Academy of Science of Turin. On 11 February, he had already given the Academy a copy of the *Introduction* [66] and on 25 February he gave Numbers 1 and 2 of Volume 4 of the *Rivista,* as well as presenting a paper of Burali-Forti, in which extensive use was made of the symbolism of mathematical logic. In fact, Burali-Forti writes, perhaps too optimistically: "The notations of mathematical logic are sufficiently known today that I may pass over an explantation of the meaning of the symbols." On 17 June, Burali-Forti gave to the Academy, through Peano, a copy of his book on mathematical logic. This work resulted from a course of lectures he had given during the academic year at the university. In the fall he became Peano's university assistant, replacing Giovanni Vailati, who became assistant in projective geometry the following year and later honorary assistant to Vito Volterra. On 18 November, Peano presented a paper to the Academy dealing with the theory of limits [70], in which very extensive use is made of his symbolism.

The publication of the first volume of H. Grassmann's collected works was the occasion for a review [69] that briefly traces the history of Grassmann's *Ausdehnungslehre*. Similarly, in a review [76] early in 1895 of F. Castellano's textbook of rational mechanics (reflecting his lessons at the Military Academy), Peano again traces the history of the vectorial method used in this book, which he would later say was the first text in rational mechanics to systematically use vectorial methods. It need hardly be added that he has high praise for it.

There were two more publications in 1894, both replies to questions posed in the *Intermédiaire des Mathématiciens* by 'Ignarus' and 'Milch', respectively. The first [70.1] calls attention to the fact that the solution given earlier by Léon Lecornu was not sufficiently rigorous and indicates that he (Peano) gave the correct solution in a note on Wronskians in 1889 [17]. Lecornu, who had been shown Peano's comment, had a ready excuse, but admitted that Peano was right.

The second [70.2], which expresses the sum of an infinite series by a definite integral (in the manner of M.A. Parseval), uses Peano's symbolism to do so. The editor, fearing the probability that "the result, in this form, will tell many of our readers nothing," gives a translation into ordinary language. But he is sympathetic, and recommends Peano's works. In all, Peano wrote fifteen times to the *Intermédiaire des Mathématiciens*, twice with questions and thirteen times with answers.

This year ended as it had begun — with student disorders. The provocation this time was the announcement of a rise in university fees. The following proclamation of the rector, Luigi Mattirolo, dated 15 December 1894, tells the story:

Students!
The disorders that have taken place in recent days must absolutely be censured.
No matter what the complaint, if made with rioting and violence, it is illegitimate and cannot be effective.
Neither you, nor any other citizens, are lacking the means to legally present petitions and observations that you think opportune: it will always be my concern to immediately take them to the competent authorities, and to support them when they appear to me to be just.
The cause that has occasioned your agitation of recent days has no foundation in fact, as appears from the following telegram of His Excellency the Minister.

You may assure the students that the rise in fees does not pertain to students who are already enrolled in any year; this will apply only to those who enroll after the promulgation of the law. Those who are presently enrolled may continue and finish their studies according to the norms now in effect, in accordance with Article 10 of this very law.

Next Monday, the 17th of this month, the university will reopen. I have firm faith that order will no longer be disturbed, and that I shall not be compelled to fulfill my burdensome duty of using all means that the law gives me to repress disorders and maintain complete freedom for study.

In 1895 Peano returned to the concept of the integral in an article [78], in which he gave a brief historical summary and compared his definition with those of other authors.

The year 1895 began and ended with official recognition of Peano's merit. On 28 February he was made a *Cavaliere* of the order of the *Corona d'Italia* (Knight of the Crown of Italy.) This order was established on 20 February, 1868, to commemorate the independence and unity of Italy (and celebrate the annexation of Veneto.) It was commonly given in recognition of merit, to both Italians and non-Italians, the first crosses of the order being distributed on the occasion of the marriage of Prince Umberto of Savoy with Princess Margherita of Savoy-Genoa on 22 April, 1868.

The year ended with Peano's promotion to Ordinary Professor, by decree of 15 December, but effective as of 1 December, 1895. In the meanwhile there occurred the most bitter scientific controversy of his career. His opponent was Vito Volterra and their arena included both the Academy of Sciences of Turin and the Academy of the *Lincei* in Rome. That story must now be told.

CHAPTER 8

THE CONTROVERSY WITH VOLTERRA

On 29 October, 1894, Etienne Jules Marey, a distinguished physiologist and pioneer in motion photography, presented to the Academy of Sciences of Paris a sequence of 32 photographs he had made of the fall of a cat dropped with feet upward. They showed that the cat completed exactly a half turn. This seemed to be in contradiction to the principle in mechanics of the conservation of angular momentum, and gave rise to a discussion in which several members of the Academy and, later, even the daily newspapers took part, several explanations being given of how the cat was able to turn over. In the January 1895 issue of the *Rivista di Matematica*, Peano gave another explanation in 'The principle of the conservation of angular momentum and the story of a cat' [77]. After mentioning several other possible explanations, he wrote:

But the explanation of the cat's motion appears to me quite simple. When the animal is left to itself, it describes with its tail a circle in the plane perpendicular to the axis of its body. Consequently, by the principle of the conservation of angular momentum, the rest of its body must rotate in the sense opposite to the tail. When it has rotated as much as it wishes, it halts its tail and with this simultaneously stops its rotary motion, saving in this way itself and the principle of angular momentum.

He goes on to say that this motion can be detected in the photographs and even be seen by the naked eye. It is not clear how serious Peano is about this explanation; this article has the heading 'Variety' and he even makes a joke about doing the experiment with a trustworthy cat. Although he probably read the account in the *Comptes Rendus* of the Academy of Sciences of Paris, he seems not to have had a copy before him when he was writing. (He refers to Marey, for example, as "Murley".)

Maurice Lévy, at the same session of the Academy, pointed out that a cat *could* turn over by rotating head, or neck, or tail, or feet, without suggesting that this cat *did*. In the photographs it appears that the cat's tail does revolve in the direction opposite to the rotation of its body, but only one full turn. The explanation of Guyou, with which Marey concurred, was closer to more recent explanations, namely that the cat first turns its front end and then its rear, and is able to do this by bringing the part to be turned closer to the axis of rotation, thus decreasing the moment of inertia of that part.

At the next session of the Academy on 4 November 1894, Marcel Duprez presented an apparatus that exhibited how internal motions of a body could change its orientation and Paul Appell gave a few remarks on mechanics, that were suggested by all this.

On 5 May, 1895, Peano presented a note to the Academy of Sciences of Turin [79] that begins by recalling the above events, and then says:

This naturally presents the question: Can our earth change its own orientation in space, using only internal actions, just as every other living being? From the viewpoint of mechanics, the question is identical. But Prof. Volterra has the merit of having proposed it first. He made it the object of several notes presented to this Academy, of which the first is dated the 3rd of last February.

Peano then shows how the geometrical calculus can be used to study this possibility. He takes as his example the action of the Gulf Stream and, without exhibiting his calculations, gives the conclusion that it alone would displace the North Pole approximately 3 mm a day, or about 1.1 meters a year. He insists that this is only a very rough approximation.

The implication that Volterra's work was inspired by Peano must have irritated him greatly, and Peano, by presenting results of calculations on the Gulf Stream that 'stole Volterra's thunder', added injury to insult. His reaction was immediate. Peano's calculations, he said, started from ideas that he himself had already explained in his various notes, but were based on rather unreliable data. In his notes and in a memoir sent long ago (it was, however, dated 1 February, 1895) to the *Astronomische Nachrichten* he had treated this question. As for numerical applications based on real, rigorous data, he had already some time ago carried them out, starting from the important results of Chandler, etc., but was keeping them for presentation to the Academy after he had developed the general theory, which he planned to do by a note this session, but he could even give his note 'On the periodic movements of the earth's pole' also, if allowed a bit more time. He was and did.

Vito Volterra, who was called to the University of Turin as Ordinary Professor of Rational Mechanics in 1893, was elected a resident member of the Academy of Sciences of Turin on 3 February, 1895, the date he presented his first paper on this subject. This had the same title, 'On the theory of the movements of the earth's pole', as the memoir sent to the *Astronomische Nachrichten*. Although the topic was, to a certain extent, 'in the air', he could say in the memoir: "As far as I know, there has up to now been no examination of the action (which may be strong) that internal forces exert on the positions of the earth's poles, on their movements, and consequently on the variations in latitude."

The question was indeed current. Euler had shown in 1765 that in a flattened sphere of the form and dimension of the earth, which rotates about an axis not quite coincident with the axis of symmetry, the axis of rotation must appear to describe a narrow cone every 305 days, and consequently the latitude of all places must show small periodic changes, but these changes were not detected until F. Küstner, in 1889 and 1890, made continuous measurements of the latitudes of Berlin, Potsdam, and Prague. S.C. Chandler also discovered in 1891 that each pole revolves about its mean position with a radius of about thirty feet in a period of 427 days. In 1892, Simon Newcomb explained the discrepancy between Euler's and Chandler's observations by the theory of the earth's elasticity.

Following up his paper of 3 February, in which he had said, "for example, ocean currents could in a certain way be regarded as movements of the type considered here," Volterra presented two more notes, on 3 March and 31 March, before the 5 May session at which the two notes mentioned earler were presented. Peano presented a paper on 19 May, which he withdrew before publication, having found an error in it. (He wrote to Volterra to tell him this.) Volterra presented his sixth and seventh notes on this subject on 9 June and 23 June, respectively.

Peano followed Volterra's presentation on 23 June with his second note on this subject, in which he poses the question of finding the position of the pole after a certain time, given the existence of the internal motion of the earch [80]. The treatment was theoretic and geometrical calculus was used. Peano begins by recalling the work of Zanotti-Bianco, Eneström, Gessel, Gylden, Resal, W. Thomson, G. Darwin — there is no mention of Volterra! This was too much; Volterra now moved to another forum, presenting his next note on the subject to the Academy of the *Lincei* in Rome (of which he was a corresponding member) — and he did not fail to mention Peano. He begins:

Prof. Peano in a note presented to the Academy of Turin in the session of 23 June of this year, and which has just now been printed, shows that a system which is symmetric about an axis and which constantly maintains its form and density distribution, may have variable internal movements that follow a law such that the rotational pole moves continually farther from the inertial pole.

Seeing that this result can be obtained as an evident and immediate consequence of formulas considered by me and explained-in several preceding memoirs, which Prof. Peano forgot to cite, although they were published this year in the same Acts of the Academy of Turin, I may be allowed to show this here avoiding the employment, made by the said author, of methods and notations not generally accepted, and of proceedings hardly suited to making clear the path taken and the result reached.

The 'methods and notations not generally accepted' were, of course, those of the geometrical calculus, but he was probably also irritated by Peano's apparent reference to the simplification he had made in his equations. (Peano had written: "To decide such a question it is necessary to make complete calculations, omitting nothing.") Volterra was naturally angered by Peano's obviously deliberate refusal to cite him.

Peano accepted the challenge, and in a note to the *Lincei*, presented by Beltrami, needled his opponent with the very title, 'On the motion of a system in which there are variable internal movements' [84], that Volterra had used. He wrote:

In an article having this same title, and published in the Proceedings of the Royal Academy of the *Lincei* last 15 September, Prof. Volterra confirmed with his calculations one of the results that I reached in my two notes 'On the displacement of the earth's pole,' published in the Acts of the Royal Academy of Turin, under the dates 5 May and 23 June. And since the question of the pole's motion is of interest now, I believe it useful to explain in a few words the results which I reached, and which can be found by either way.

He then mentions the cat experiment and his January note – one almost hears Volterra exclaim, "What does his damn cat have to do with it!" – and says:

Later, discussing the question of the displacement of the earth's pole, produced by movements of parts of the earth, such as the ocean currents, I pointed out to several people the identity of the two questions, seeing that instead of a cat and its tail, one could talk about the earth and its ocean. But as no one agreed with me, I published my first note (5 May). . . .

Now, if the question can be easily and completely solved by my way, the reason others run into great difficulties, I believe, depends on their habitual use of long formulas to indicate simple ideas.

Thus Prof. Volterra, in his memoir 'On the theory of movements of the earth's pole (Acts of the Academy of Turin, 3 February, 1895), began by writing three equations that, understood geometrically, say that a certain vector is constant. He differentiates them, transforms them, and in the general case which interests us here, arrives at a single integral (on the last page of the memoir), which signifies "The length of this vector is constant."

After more of the same, he concludes: "I believe it useless to add anything else, seeing that finally Prof. Volterra agrees with my result that 'relative movements, however small, acting for a sufficient time, can displace the earth's pole, even supposing the continents rigid'."

In exasperation, Volterra wrote a letter to the president of the *Lincei* on 1 January, 1896, which President Brioschi communicated to the Academy on 5 January. It begins:

Dear Mr. President:

Please allow me to communicate to you a brief reply to the note of Prof. Peano ...

Relative to what is found said in the beginning of his note, it seems to me that it is not worth the effort of spending any words, seeing that no one can doubt my priority, whether with respect to treating the question, or with respect to the fundamental idea which forms its point of departure; nor can any doubt arise about the originality of my idea, as I explained in my lectures of last year, which I found while searching for an apt example to illustrate Hertz' concept of substituting hidden movements for the consideration of forces in the investigation of natural phenomena; and it is not necessary for me to justify myself with the cat question, as Peano hints, a question, for that matter, about which he limited himself to writing in his journal a simple and brief review of the work of others ....

Thus having announced this conclusion, which was immediate and evident from my considerations, without citing me, but only dressing it up in vectorial language, Peano deserves the censure contained in the note I presented to the Academy last September. And so, I have no need of agreeing with any of Peano's results.

And he concludes:

Having thus shown to be empty and unfounded any of the points of criticism made of me by Peano, and that his assertions are neither original nor exact, he himself having recognized them as such, for my part I hold this polemic definitively closed.

As Volterra must have anticipated by now, Peano would not allow him to have the final word, and on 1 March, 1896, Brioschi presented to the *Lincei* Peano's last note on this subject [89], having the same title, 'On the motion of the earth's pole,' as the 5 May note of the previous year. Surprisingly, however, he does not mention Volterra. He justifies the note in the first paragraph:

In my note, published in the Acts of the Royal Academy of Turin, 5 May 1895, and having this same title, after applying the *Ausdehnungslehre* of H. Grassmann to the principles of mechanics, as a numerical exercise of the last theorem, I proposed to estimate the velocity with which the polar regions are being displaced by virtue of the motions of the fluid parts of our globe. But having made use of this new method, and having in a note 'On the motion of a system in which there are internal motions' (Proceedings of the Royal Academy of the *Lincei*, 1 December, 1895) explained my results without proof, doubt could remain on the part of some readers. Hence it will not be completely inopportune to translate some of these formulas into cartesian coordinates. The formulas become longer, but the reasoning loses none of its simplicity.

He briefly carries out his plan, and then says in the last paragraph:

The conclusion is that the intensity of the actual atmospheric and marine currents is more than sufficient to give to the poles vast irregular movements, of arbitrary size, and this whether one supposes the earth to be plastic, as Schiaparelli affirms, or to be rigorously rigid. It is up to astronomy and geology to recognize whether these displacements are taking place, or have taken place.

A footnote to this polemic was written in 1958, when Ugo Cassina, using the 'Italian' vectorial calculus, re-worked the problem presented by Peano in his 5 May and 23 June, 1895, notes ([79] and [80]). Using the rough data of Peano, Cassina's calculations led to a result substantially in agreement with that of Peano (who did not present his calculations.)

This controversy illuminates the scientific personalities of the two antagonists. Volterra was careful and methodical, a master of classical analysis, especially elliptic functions. He displayed all his work, carefully and thoroughly, and having announced the study of a new field, expected its investigation to be left entirely to him. Peano was the innovator, eager to show the power of his geometrical calculus, who delighted in applications that caught the imagination and was not above using homely examples. (One such is the note dated 24 November, 1895, but published in the 1896 volume of the *Rendiconti del Circolo Matematico di Palermo* [87], in which he explains how the action of a child in a swing can increase the amplitude of the oscillations.) Furthermore, he believed in scientific collaboration, as his work on the *Formulario* clearly showed.

Of course, there was probably more to this controversy than is contained in the documents. It is likely that Volterra resented Peano's insistence on using the symbolism of his mathematical logic in the classroom, as well as his propaganda for the *Formulario*. This no doubt accentuated the rift that perennially seems to exist, on university faculties, between professors of pure and professors of applied mathematics. So this argument with Volterra may have been more symptomatic of their differences than the cause. Certainly the situation did not improve, and Peano's estrangement from other members of the mathematics faculty continued, even after Volterra's transfer to Rome in 1900.

The year 1895 is also notable because of Peano's (happily, non-controversial) contact with Georg Cantor. The documentary evidence of this is far from complete; the correspondence in both directions has apparently been lost. What we do have, however, are the 'letter books' of Cantor, which contain drafts of 10 letters from Cantor to Peano during the period from 6 April, 1895, to 10 January, 1896, and give evidence of at least 8 letters and postcards from Peano.

The letter books show that Cantor was in correspondence with Giulio

Vivanti (at that time in Mantua) as early as 1886, and it was a letter to him in December 1893 that led to the correspondence with Peano. That letter closed with the lines: "With my friendliest greetings to Mr. Peano, to whom I hope you will show this letter." Nevertheless, the indication that this was done came only in Cantor's letter of 6 April, 1895, in which he thanks Peano for a copy of the *Formulaire* and a postcard in which Peano had asked for permission to publish the letter to Vivanti in his journal. Cantor wrote giving this permission, but asked that the personal comments be "toned down." (In fact, Peano left them out when part of the letter was published in the *Rivista di Matematica* in August of that year.) Cantor also remarks that he had sent the first two articles of his 'Beiträge zur Begründung der transfiniten Mengenlehre' to the *Mathematische Annalen*. In the next letter, 27 July, 1895, Cantor says he would be pleased if Peano would publish an Italian translation of the 'Beiträge', offers to send later on an original article on 'infinitesimi', and asks for Peano's opinion of Veronese. Cantor says regarding Veronese: "It is difficult to grasp a bull by the horns, when the horns are missing. By the horns being missing, I mean here the lack of clear explanations of what he means." Without waiting for Peano's reply – in fact, the very next day – Cantor composed a letter to Peano that he asked to be published, and it was also published in the August issue, following the letter to Vivanti. This letter contains comments on the German version of Veronese's book on the foundations of geometry that was published in 1894. Although Cantor did not wait for Peano's opinion, there is no doubt that Peano would have agreed with his remark about the "bull's horns".

On 8 August, 1895, Cantor wrote Peano in reply to Peano's letter of 4 August. From it we learn that Peano was trying to understand Cantor's ordinal types. Cantor, for his part, had not yet read Peano's *Formulaire*, but he promised to take it with him on his holiday in Westerland on Sylt. He says that he will send the next day a copy of the proof of his first article of the 'Beiträge' and once again remarks that he would be pleased if Peano publishes an Italian translation. Peano sent a long letter commenting on the 'Beiträge', to which Cantor answered on 14 September. On 18 September, Peano wrote the news that Francesco Gerbaldi would translate the 'Beiträge'. And Peano, apparently having had enough of the transfinite numbers, asked: "Where does one find a definition of finite cardinal numbers?" Cantor's reply was: $1 = \overline{(a)}$, $2 = \overline{(a,b)}$, $3 = \overline{(a,b,c)}$, ...

Already on 3 November Peano could write Cantor that the first article of his 'Beiträge' was translated and would soon be sent for proofreading. Cantor replied on 9 November that he would write an addition to it for Peano's journal. This was not done, or at least was not published. Probably Peano surprised him by sending the proofs of the first 'Beiträge' article so quickly: Cantor wrote again on 29 November to say that he was sending the proofs to Peano's printer that same day. He notes that Friedrich Meyer, whose Italian was better than his, found the translation 'outstanding'. It was published in the November issue of the journal and Cantor wrote Peano on 6 December, 1895, to ask for 100 offprints. Thus the Italian translation of the first of the two 'Beiträge' appeared the same year as the original German was published in the *Mathematische Annalen*.

The last letter-draft is dated 10 January, 1896, and answers a letter of Peano of 18 December. He asks for a photograph of Peano (one of Cantor's preoccupations – he must surely have had one of the best collections of photographs of contemporary mathematicians!) and hopes that Peano has recovered from the "severe illness (pleuritis)." This last remark is one of the rare indications of an illness of Peano.

CHAPTER 9

THE FIRST INTERNATIONAL CONGRESS OF
MATHEMATICIANS

From the beginning, the first volume of the *Formulario* was meant only as a first approach to he comprehensive list of formulas that Peano hoped to publish. He knew the value of collaboration and expected that readers of the *Formulaire de Mathématiques* would send him their corrections and additions. Many of them did, from all over Europe, but mostly the contributors were Italian, especially those who had personal contact with Peano in Turin – not including, however, other members of the university mathematics faculty.

Continued work on the *Formulario* now concentrated on mathematical logic, and resulted in the separate publication of Part 1 of Volume 2 of the *Formulaire* in 1897 [93]. This was devoted exclusively to mathematical logic (Parts 2 and 3 were published in 1898 and 1899, respectively), and was dated 11 August, 1897, the day Peano 'presented' it to the First International Congress of Mathematicians in Zurich. We now discuss Peano's work leading up to this event.

After the publication of the first volume of the *Formulario* early in 1895, there were only a few other publications of Peano that year other than those of the polemic with Volterra, already discussed. A letter to the editor of the *Monatshefte der Mathematik* (Vienna) [81] seems less concerned with asserting priority to a theorem contained in the *Formulaire*, than with stating the value of the *Formulaire*. The theorem in question deals with Peano's proposed definition of 'limit' of a function (discussed in [52]), the bounds of which are called here 'limits of indetermination'. The review [82] of one of Frege's publications will be discussed later.

Peano's continued interest in the "geometrical calculus according to the *Ausdehnungslehre* of H. Grassmann" is shown in two notes. One is a bibliography of works directly related to the *Ausdehnungslehre* [83], that includes 70 works of 29 authors, including 7 publications of Peano, the last being the note 'On the motion of the earth's pole' [79], presented to the Academy of Sciences of Turin on 5 May, 1895. The date '6 June' is also given, as if this note were presented in two parts. (This cannot be correct, since 6 June was a Thursday, and the Academy always met on Sundays.) This probably refers to the note presented on 19 May and then withdrawn. Presumably Peano had planned to present a corrected version, possibly at the 9 June session which,

64

in fact, he did not attend. The other note dealt with the geometrical calculus ('Linear transformations of vectors in a plane' [85]) and was presented to the Academy on 1 December, when he also reported on the memoir of F. Giudice that he (Peano) had offered at the preceding session, and which he and D'Ovidio had been asked to review [86]. (They recommended publication.)

There was one other publication in 1895, a reply to a question in the *Intermédiaire des Mathématiciens* asking for the correct definition of the 'curve of Agnesi' [86.1]. Peano must have had a copy of Maria Gaetana Agnesi's book, since he quotes the definition verbatim. (The typographical error in his statement is most likely due to a faulty justification of the right-hand margin.) He gives a reference to a textbook of Francesco D'Arcais and notes that the curve is "rather called the *versiera*," having apparently realized the double mistake he had made in his *Applicazioni geometriche* of 1887, where he defined a curve that is not the *versiera* and said it was called the "*visiera* of Agnesi"! Much later, Gino Loria wrote of this curve: "It is thus a curve quite distinct from the *versiera* and so may keep the name *visiera* given it by Peano."

Early in 1896 Peano announced the preparation of the second volume of the *Formulario* in the first four pages of Volume 6 of the *Rivista*, in a note entitled 'Introduction to Vol. 2 of the *Formulaire de Mathématiques*' [88]. He has now changed the name of the journal to the French *Revue de Mathématiques*. This change is symbolic of the growing importance of the *Formulario* project, especially in relation to the journal. The first five volumes of the *Rivista* had each covered a calendar year. During this period the *Formulario* was begun as an auxiliary project. With Vol. 6 the roles are becoming reversed, and the journal's last three volumes (6–8) each covered several years, terminating in 1906, the year the publication of the fifth and final volume of the *Formulario* began. What began as an auxiliary project became a case of the 'tail wagging the dog.' There would be one more change in the title of the journal, from French to *Latino sine flexione*, again reflecting the language used in the *Formulario*.

In this 'Introduction' he traces the beginning of mathematical logic to Leibniz, giving several quotations, including Leibniz' remark: 'I spoke of my *specieuse générale* to the Marquis de l'Hospital and others, but they paid no more attention to me than if I had narrated a dream.' Peano adds:

After two centuries, this 'dream' of the inventor of the infinitesimal calculus has become a reality. . . . We now have the solution to the problem proposed by Leibniz. I say 'the

solution' and not 'a solution', for it is unique. Mathematical logic, the new science resulting from this research, has for its object the properties of the operations and relations of
logic. Its object, then, is a set of truths, not conventions.

After describing the advantages of the *Formulario*, he makes the ominous
statement: "Each professor will be able to adopt this *Formulario* as a textbook, for it ought to contain all theorems and all methods. His teaching will
be reduced to showing how to read the formulas, and to indicating to the
students the theorems that he wishes to explain in his course." True to his
word, as the parts of the *Formulario* relating to the calculus were published,
Peano immediately began using them in his classes, with the unfortunate
consequences that will be described later.

Political events were the cause of brief student disorders in March, when
the Italian forces in Abyssinia met final disaster in the battle of Adowa on
1 March, 1896, in which 4500 men and 2 generals were killed, 2000 wounded,
and 1500 taken prisoners. The rector of the university relates:

During this academic year the order and discipline of our schools were disturbed only
momentarily in March 1896. The unexpected news of the disastrous battle of Adowa,
which grieved all Italian souls, was the cause of it. Reading the accounts giving a narra-
tion of disorders in other cities, especially in Milan and Rome, agitated our students and
shook them out of the barely checked calm in which they remained during the days of
Tuesday the 3rd and Wednesday the 4th. The agitation, however, incited also by elements
outside the university, as always happens, stopped when our illustrious predecessor in a
forceful speech made them understand that the discipline and the order of the school are
and must be entrusted especially to the students themselves. When his student audience
understood their duty, there was a lively reaction against the rioters, and shortly after-
ward order was perfectly reestablished again. Nor were there further disturbances, despite
the fact of conflicting news concerning the unfortunate events in Africa.

As a result of the wave of national grief and anger that swept the country,
Francesco Crispi, the prime minister, resigned his second ministry, which had
begun in 1893.

In the meantime, Peano says he was urged to give a succinct introduction
to the geometrical calculus. It is possible that he was criticized by members
of the Academy of Sciences for using it and wished to propagandize. At any
rate, soon after returning from a trip to Rome (from 16 May to 2 June, 1896)
he presented to the Academy, on 21 June, an 'Essay on the geometrical calculus' [90], addressed "not to students, but to colleagues." This excellent
article was translated into Polish by Samuel Dickstein (1897) and into German by A. Lanner (1897–98).

In July, Peano went on a hiking expedition on the slopes of Monte Rosa.

During this year, Peano moved to a fourth-floor apartment at Via Barbaroux, No. 4, a very central location, close to the university and only a few blocks from the Royal Theater, where the premiere of Puccini's 'Boheme' took place on 1 February. Two years later he moved to No. 6 of the same street, but by the fall of 1903 he is back at the No. 4 address, where he would remain until his death in 1932, spending his vacation periods, however, at his villa in Cavoretto. In his modest Via Barbaroux apartment he never had a telephone or stove; it was heated by an open fireplace. His desk was a small table. With his passion for dogs, he probably kept at least one there; he always had several in Cavoretto.

In the fall, Angelo Ramorino became Peano's assistant at the university. He was also helpful in the *Formulario* project, contributing a number of additions and corrections. Vailati is now 'voluntary assistant' for the third-year mathematics course.

Peano's five remaining publications of 1896 were all answers to questions in the *Intermédiaire des Mathématiciens* [90.1−5]. Camille Jordan had asked: "Can anyone indicate a curve, $x = f(t)$, $y = \phi(t)$ (f and ϕ being continuous functions) whose area is indeterminant?" The question is ambiguous; Peano's answer is not, although it probably did not answer the question Jordan intended. Peano says, in effect [90.1]: Take my curve that fills a square and connect the end points with an ordinary curve. Then the least upper bound of the areas of the interior polygons is less than the greatest lower bound of the areas of the exterior polygons, and that by an amount that is exactly the area of the square.

It was a question of 'Lausbrachter', however, that was most welcome to Peano: "What is the most general question that has been posed or resolved up to now in the whole of the mathematical sciences?" [90.5] By now, we almost anticipate, in his answer, the opening word 'Leibniz'! After quoting Leibniz' description of the *specieuse générale*, "a kind of universal language or writing, in which all the truths of reasoning will be reduced to a sort of calculation," Peano says: "According to Leibniz, then, the most general question that has been posed is the construction of this *specieuse*." And he repeats what he had said earlier that year in the Intoduction to Vol. 2 of the *Formulaire de Mathématiques*: "I published, in 1889, for the first time, a small book, written entirely in the symbols of mathematical logic." He closes with a brief description of the *Formulario* project and invites correspondence.

Early in 1897 there is another round of student disorders, which required sterner measures. Here is the description of Domenico Tibone, the rector:

Unfortunately this academic year the order and discipline of the schools was disturbed very strongly. The news of the disorders that took place at the University of Bologna gave an unjustified excuse to a small part of our students to rise up, so to speak, and proclaim themselves in the name of all, even those who had never thought about it, to be in solidarity with the students in Bologna. They found their pretext in a telegram that the President of the University Monarchical Club sent to His Excellency the Minister of Public Instruction, stigmatizing the way in which he was received by these students. The disorders began at two o'clock on the afternoon of 3 February, the work of a few, and grew so much in the following days that it necessitated the closing of the university, which lasted from 8 February until 4 March. In the meantime, the principal agitators having been uncovered, I made a formal accusation against them, on the basis of which legal proceedings began that terminated with the absolution of several and with the sentencing of one to temporary expulsion from the university for the entire academic year, of five to suspension from the summer and fall examination periods, of six from just the summer exams. Two were given a simple admonition.

This forceful action had a good result, for on the reopening of the university, all the students returned to their duties and there were no more disturbances for the remainder of the academic year.

The 'Mathesis' Italian Society of Mathematicians, an organization of secondary school teachers of mathematics, was organized in 1895, and by early 1897 Peano, although not a member, participated in the meetings of the local chapter in Turin. It is recorded in the minutes of the 28 February 1897 meeting that both he and Alessandro Padoa took part in several discussions. Padoa, an 1895 graduate of the University of Turin, while he did not actually study under Peano, became an ardent disciple of Peano and champion of his mathematical logic. Although primarily a teacher in the secondary schools, he gave a series of lectures on mathematical logic at several universities, beginning with the New University of Brussels in 1898. The first national congress of the association would be held in Turin in September 1898, and Peano would take a very active part.

On 4 April, Peano presented to the Academy of Sciences of Turin his 'Studies in mathematical logic' [91]. In this work he is concerned with reducing the number of undefined terms to a minimum. The theory is here based on the following seven conventions: (1) the variables $a,b,c, \ldots ,x,y,z$; (2) parentheses and dots to separate parts of a formula; (3) K for 'class' (later changed to 'Cls'); (4) ϵ for set membership; (5) $\supset_{x, \ldots ,z}$ for the universally quantified conditional; (6) $\cap$ (or juxtaposition) for conjunction; and (7) $(x;y)$ for the ordered pair formed of x and y. With regard to this last, he specifically remarks: "The idea of an ordered pair is fundamental, i. e. we do not know how to express it by the preceding symbols." (This would be done by

Norbert Wiener in 1914.) It was in this article that he introduced the symbol Ǝ for existence, i. e. if a is a class, then ' Ǝ a' means 'the class a is not empty'. This paper also contains remarks on Frege's mathematical logic, which will be discussed in the next chapter. A German translation of this article was published in 1899 as an appendix to the German translation of the *Calcolo differenziale* of 1884.

Publication of the *Rivista di Matematica* was slow this year, so that by 10 April, 1897, only the first fascicle of Vol. 6 had appeared. Peano explained in a postcard of that date to J. Lüroth that the delay was due to additions and corrections to Vol. 2 of the *Formulaire*, a part of which was to appear along with the second fascicle.

On 20 June, S. Pincherle presented Peano's note 'On Wronskian determinants' [92] to the Academy of the *Lincei* in Rome. He first states the theorem: If several functions are linearly independent on an interval, then their Wronskian is identically zero. After giving a counter-example to show that the converse is not true (which he first gave in 1889 [17]), he then states and proves a partial converse. Peano's efforts next concentrated on getting Part 1 'Mathematical Logic', of Vol. 2 of the *Formulaire*, printed in time for the International Congress of Mathematicians in August.

For several years there had been talk of a possible international congress. Indeed, as early as 27 July, 1888, Georg Cantor had written:

Would it not be possible in some neutral place such as Belgium or Switzerland or Holland to hold a gathering of French and German mathematicians in the near future? I remember that, even in more difficult times, in the 70's, the cordial friendship between Mr. Hermite and the immortal Mr. Borchardt served admirably as a strong bond holding the scientific community above the struggle of the nations. This splendid example should not be lost from sight. [From a letter dated 27 July, 1888, from Halle a. d. Saale (catalogued in the E.T.H. Library, Zurich, as 'Letter to an unnamed Parisian colleague').]

Finally, at the suggestion of Heinrich Weber in Strasbourg and Felix Klein in Göttingen, and with the agreement of the Deutsche Mathematikervereinigung and the Société Mathématique de France, Geiser sent a circular letter on 16 July 1896 to the mathematicians of Zurich calling for a meeting to discuss this. At the first meeting on 21 July an organizational committee was named, composed of Professors Geiser (as president), Ferdinand Rudio, Adolf Hurwitz, A.J. Franel, Weber, and the Assistants Rebstein (who served as German secretary) and Dumas (who served as French secretary.) At least nine more official meetings were held. In January 1897, the first announcement of a three-day congress, to be held in Zurich, 9–11 August 1897, was sent to some 2000 mathematicians and mathematical physicists. The international committee who signed it included: Luigi Cremona (Italy), Felix Klein (Germany), Henri Poincaré (France), and George William Hill (U.S.A.), whose research in lunar theory began, in 1877, the study of infinite determinants.

In February, a second notice was sent out, calling for papers and suggestions. On 18 February, Geiser announced to the committee that Poincaré and Aurel Boleslav Stodola would lecture and on 11 May that the lecturers for the general sessions would be men of repute [Männer, deren Namen einen Klang haben]: Klein, Poincaré, Hurwitz, Stodola. But Stodola was now questionable because of his health, and before the end of May the final announcement of the congress was sent out, that included a provisional program and in it Peano has replaced Stodola.

The above account of the origin of the First International Congress of Mathematicians is taken from material in the Archives of the E.T.H. Zurich. This is certainly not the whole story. The statement of H. Meschkowski that Cantor "pressed for international congresses of mathematicians and was responsible for bringing about the first ever held, in Zurich in 1897" [*Dictionary of Scientific Biography* 3, 53] is perhaps an exaggeration, but it should be noted here that a 2-page propaganda leaflet (dated Paris, 3 August 1897) for the journal *Intermédiaire des Mathématiciens* was distributed, in which it was claimed that the editors (C.A. Laisant and Emile Lemoine) were the originators of the idea. Meschkowski was perhaps referring to a letter Cantor wrote Laisant from Halle on 19 March 1896, in which he said that he wanted the three of them (Cantor, Laisant, and Lemoine) to begin to work out the international organization of mathematicians that was supposed to be set up in Zürich in 1897, and he urged Laisant to keep this a secret among themselves. Certainly there were many undercurrents.

The congress opened the evening of Sunday 8 August with a reception. In all there were 242 delegates to the congress (of which 38 were women) and 16 countries were represented: 68 from Switzerland, 53 from Germany, 32 from France, 25 from Italy, 20 from Austria-Hungary, 13 from Russia, and 7 from the United States. The American delegation was made up of: William Fogg Osgood (Harvard), James Pierpont (Yale), Charlotte Angas Scott (Bryn Mawr), James Harkness (Bryn Mawr), René de Saussure (Catholic Univ.), and two students, C.L. Bouton and G.R. Ohlhausen, both of St. Louis. Five of the Italians were from Turin. Besides Peano, there were Corrado Segre, Vito Volterra, C. Burali-Forti, and Giovanni Vacca, who had just graduated from the University of Genoa and had been selected by Peano to be his assistant in the fall. Ulrico Hoepli, the publisher, was there from Milan with a book display that included Burali-Forti's manual of mathematical logic. Peano, of course, made sure that copies of the *Formulaire de Mathématiques* were available.

The first general session of the congress took place on Monday morning, 9 August. After the welcoming speech, there were papers by H. Poincaré and A. Hurwitz. (Prof. Franel read the paper of Poincaré, who was unable to attend the congress.) Dinner was scheduled for 1:00 p.m., to be followed by a boatride on the Lake of Zurich at 4:00. The return trip, at 9:00, was the occasion for a 'Venetian Night.'

On Tuesday the congress divided into five sections for the presentation of 30-minute papers. Peano was vice-president of Section 1, 'Arithmetic and Algebra', and presided over the afternoon session, at which five papers were read, three others having to be cancelled for lack of time. Peano was most interested in the talk of Ernst Schröder 'On pasigraphy, its present condition and the pasigraphic movement in Italy.' Schröder had a great deal of praise for Peano and his school, although he found Peano's statement that "the problem has now been solved" to be a bit "sanguine". (Peano published a brief review of this paper in the *Rivista di Mathematica* in the fall of 1898 [102].)

In Section 3, 'Geometry', Burali-Forti's paper on 'The postulates for the geometry of Euclid and of Lobachevskii' gave rise to a discussion in which Schur and Schoenflies took part.

Peano and Klein spoke at the second general session on the morning of Wednesday, 11 August. As Peano did not actually read the paper he 'presented' [93], the Proceedings of the congress gives only the following abstract (written by Peano) of what took place:

The author gives complimentary copies of his booklet 'Mathematical logic', which constitutes Section 1 of Volume 2 of the *Formulaire de Mathématiques*, to the members of the congress.

He mentions the long sequence of philosophers and mathematicians, from Aristotle and Leibniz up to our day (page 18 of the booklet), to whom we owe the development of this new science.

The quickest way to form a clear idea of mathematical logic, its scope, its applications, and related matter, is to see it in action; which is not difficult.

In fact, the study of it does not require prior studies. Rather, one begins with the mind *tabula rasa*, except for those few ideas listed on page 3, and indicated there with conventional symbols, which the author reads.

These symbols are sufficient to express every idea and every proposition, without having to recur to ordinary language.

To this end, the author invites the reading of, and reads, the principal propositions contained in the booklet.

Indeed, the reading of this booklet required some encouragement. There was no preface, and the one page of symbols and abbreviations is abruptly followed by sixteen pages of closely packed formulas. There is then one page of bibliography and 45 pages of notes. The real preface comes at the beginning of the notes and opens with (the definition?): "Mathematical logic studies the laws of logic with the aid of the formulas and methods of the mathematical sciences."

In the afternoon there were special trains to take the delegates to the nearby mountain of Uetliberg, where a banquet in the renovated Hotel Uetliberg closed the congress. The souvenir menu had a picture of the E.T.H. Zurich and portraits of the famous Swiss mathematicians Leonhard Euler, Daniel, Jakob, and Johann Bernoulli, and Jakob Steiner (who was a great-uncle of C.F. Geiser.) Emile Picard, as president of the French Mathematical Society (it had been decided to hold the next congress in Paris in 1900) gave the first toast. Klein invited everyone to the third congress in Germany. At 2:30 the trains began taking the delegates back to Zurich, but many stayed to enjoy the view. It was indeed beautiful, and the writer of the Proceedings waxed poetic:

The weather was incomparable; not the least cloud disturbed the radiant blueness of the sky. In honor of the mathematicians, the snow-clad mountains decked themselves with their most beautiful ermine. Säntis, Glärnisch, and Tödi, the Urner, Engelberger, and Bernese Oberland Alps, from Finsteraarhorn to Diablerets, vied with one another in an effort to make the day more brilliant. In complete and strong accord, the three-day symphony of mathematicians rang out its finale.

CHAPTER 10

CONTACT WITH FREGE

Gottlob Frege had already published his *Begriffsschrift, eine der arithmetis-chen nachgebildete Formelsprache des reinen Denkens* in 1879, while Peano was still a student. Jean van Heijenoort has said that "it is perhaps the most important single work ever written in logic" [Heijenoort 1967, 1]. This is hindsight, however, for the work had little impact on Frege's contemporaries, and it was only after many of the ideas contained in it had been rediscovered — by Peano, Russell, and others — that its true value came to appreciated. Frege did not attend the congress in Zurich and Schröder did not mention him in his talk, which was largely concerned with discussing the improvements over the logic of Boole made by Peano and C.S. Peirce.

Peano, on the other hand, in his paper, mentions Frege five times. Indeed, Peano was one of the very few up to this time who gave serious recognition to Frege's work. In perhaps the only review of Frege's *Begriffsschrift*, Schröder had so roundly condemned it that probably few others read it. In 1880, Frege wrote a long article in reply to Schröder's review, but it was rejected by four of the leading German journals of the period and was never published. Later, Russell's philosophy teacher, James Ward, gave him a copy of the *Begriffsschrift*, remarking that he did not know if there was anything of value in it, as he had not read it, and Russell did not read it until 1901. He facetiously remarked: "I was, I believe, the first person who ever read it." But, by 1891, Peano had read it sufficiently to discover Frege's symbolism for what Peano symbolized by $a \supset b$, since he mentions this in his article in the first volume of the *Rivista* [31]. He was probably not acquainted with Frege's work much earlier, since this is the first time he mentions Frege and in the 1889 article on the foundations of geometry [18] he remarked that a similar study for logic "is lacking to this date."

Sometime before the end of January 1894 Peano sent Frege a postcard and copies of several of his articles. This postcard and Frege's reply have been lost (as have all letters from Frege to Peano, except the one published in the *Rivista di Matematica*), but several letters from Peano have been preserved, along with two drafts of letters from Frege. In the first draft, probably written in late 1893, Frege thanks Peano for the postcard and article. He comments on Peano's notation for universal and existential quantification, noting that "in most logicians there appears to be a great confusion about the being of this existence affirmation." He also notes Peano's distinction between ϵ and $\supset$ (which Russell reversed to the symbol $\subset$ commonly used today), writing: "This distinction is overlooked by many writers, as by Mr. Dedekind in his publication *Was sind und was sollen die Zahlen?*"

This draft seems not to be complete, but there seems no reason to doubt that a similar letter was sent, for the next item in the correspondence is a letter from Peano dated 30 January, 1894, thanking Frege for sending some 'notes'. Peano says that he has had the university library buy a copy of Frege's *Grundgesetze der Arithmetik* (1893) and that he himself bought a copy of *Grundlagen der Arithmetik* "some time ago." Peano complains of difficulty in reading Frege's notation, but is optimistic that he will improve.

Peano wrote again only 11 days later (there is no evidence that Frege wrote in the meantime) and sent Frege copies of *Notations de logique mathématique* [66] and *Arithmetices principia* [16], and he asked Frege for a "critique or review" of the *Notations*. Peano said he was reading Frege's works with increasing pleasure, but he has questions, and he gives Frege three propositions in Peano's symbolism and asks Frege to translate them into his. This Frege did on the blank space in Peano's letter, and it seems likely that Frege sent them to Peano.

The next letter is dated 14 August, 1895, however, over a year later. In it Peano remarks that he sent Frege a copy of Vol. 1 of the *Formulaire* (1895) "some months ago." Presumably he had not received an acknowledgement from Frege. Peano tells Frege that he has finished reading the *Grundgesetze* and is writing a review of it. This review was published in the July 1895 issue of the *Rivista di Matematica* [82].

In his review, Peano is primarily concerned, not with the foundations of mathematics, but with Frege's ideography (*Begriffsschrift*), which, he says, is what makes the book important. He is. of course, most concerned with comparing Frege's notation with that of the *Formulario*, to the clear advantage (in his eyes) of the latter. Because Frege's notation is written in a connected way, Peano has difficulty in distinguishing the symbols, but he decides that there are five fundamental ones from which the others are derived. He then, too quickly, draws the superficial conclusion that, since the system of the *Formulario* is based on only three symbols, it represents a deeper analysis! Peano does not understand Frege's distinction between logical propositions and the rule for deriving them (language and metalanguage), and takes Frege to task for trying to justify in ordinary language the very proposition (thus Peano) from which others are to be derived. According to Peano, this is illusory; the attempt had better be made using the formal symbolism – and then its nature as a vicious circle would be apparent. Although the remark: "This book must have cost its author much labor; reading it is also quite tiring," could hardly be expected to please Frege, nonetheless Peano welcomes the work of Frege and urges him to apply his ideography to various parts of mathematics in the expection that uncertain points may be cleared up. Peano certainly does not feel that his own system is perfect. Rather, in the confidence that what they are trying to express is a set of truths, and not mere conventions, he expects that, as the difficulties are overcome, one will be able to translate from one system to another.

Peano wrote Frege on 29 October, 1895, to thank him for a couple of offprints and to ask if Frege had received the copy he sent of the review of the *Grundgesetze*. On 5 April, 1896, Peano wrote again to ask for a copy of the talk Frege had promised to give, dealing with "our symbolism."

Frege responded with a letter in September 1896, but before then had twice discussed Peano's work publicly: in a lecture in Lübeck (that was not published, however) and in an elaboration of this lecture submitted to the Society of Sciences in Leipzig on 6 July, 1896, with the title, 'On the ideography of Mr. Peano and my own.' He remarks in this paper that he intends to write Peano and so does not directly answer the criticisms in the review. He compares his work with Peano's, as contained in the *Introduction* [66] and the *Formulaire*, Vol. 1 [71]. He begins by stating his objective for forming an ideography, namely the search for the basic axioms on which the whole of mathematics is based. (He concludes that this is not Peano's intention.) This is an old question, he says, and if it has nonetheless not yet been answered, "the reason is to be seen in the logical incompleteness of our language." His ideography differs from speech in two ways: writing lasts, while speech does not; writing appears on a two-dimensional surface, while speech appears in "simply extended time." The advantage of the first of these is obvious to all; the second has the great advantage of allowing the writing to better reflect the structure of the ideas and of allowing this structure to be seen at a glance. He sees Peano's insistence on writing formulas on a single line as a "deliberate [*mutwillig*] giving up" of this advantage. He admits that Peano's symbolism is easier for the typesetter and often takes less room, but "the convenience of the typesetter is not the highest good" and these advantages, because of the lessening of clarity and the logical deficiencies, are bought at too dear a price – "at least for the goal I am pursing."

Frege finds several other points to criticize in Peano's symbolism, e. g. the use of the same symbol with different meanings in the same formula. Still, he finds that Peano has made great progress with respect to Boole, especially by the introduction of a symbol for universal quantification (although Frege considers his own method of handling this better, since Peano's symbol is always linked with some other logical symbol.)

On 29 September, 1896, Frege wrote Peano a reply to his review and asked him to publish it in the *Rivista*. This Peano did, but not until 1898, for, although No. 1 of Vol. 6 of the *Rivista* appeared in early 1896, No. 2 (containing Frege's letter) did not appear until January 1898. Peano apologized for the delay, saying that it was caused by work on the *Formulaire*, Vol. 2, Part 1. Frege objects to Peano's claim of a deeper analysis merely because he uses fewer signs. He also objects to the use of several definitions for the same sign. He goes on to show that, in fact, Peano's three signs are not adequate to define all his other signs. He says also that Peano has misunderstood some of his work, for example his use of German, Latin, and Greek letters. Peano later admitted that this was true, that he had imagined they were implicit hypotheses [see Jourdain 1912, 248].

Peano wrote on 3 October, 1896, to thank Frege for the letter and again at length on 14 October. He had in the meantime received from Frege a copy of the article requested. Peano says that he has re-read the *Begriffsschrift* and the *Grundgesetze* "with new pleasure, for I understand them even more."

A passage in this letter illustrates Peano's preoccupation with the *Formulaire*:

As you have said, my principal aim is the *Formulaire*, and not to occupy myself exclusively with logic or with a particular subject. My job is to cut up with scissors the old *Formulaire* and to put it back together with glue, inserting all the new additions and corrections that my collaborators have sent. Thus I unfortunately do not have the time to examine all points, and so I cannot get from your books all the benefit I wish.

He goes on to ask Frege to send the formulas he has discovered for inclusion in the *Formulaire*.

In his published reply [97], Peano agrees that his three symbols are inadequate, noting that the latest edition of the *Formulario* [93] contains nine undefined terms (although Peano suspects that further research will reduce this number.) He partly agrees with Frege's criticism of his definitions, but generally defends his procedure. He says that Frege's observation of the non-homogeneity of a certain formula is "exact" and that the formula has been eliminated from the *Formulaire*. Finally, he remarks: "The *Formulario* is not an individual work, but rather is becoming more and more a work of collaboration. All observations that contribute to its increase and perfecting will be gratefully accepted."

Frege drafted another letter, apparently in reply to Peano's letter (the draft is incomplete and there is no certainty that it was sent) that takes up again the multiplicity and conditionality of Peano's definitions, which he had earlier objected to. Peano defended his method of defining by saying that this was in fact what mathematicians were doing. Frege's reply was:

It is deplorable that among mathematicians there is no agreement on the principles that govern definitions. The introduction of such an agreement would be a meritorious service for a general congress of mathematicians. At present, complete anarchy reigns in this field, a fact that is doubtless convenient for authors of mathematics, but prejudicial to science.

We shall see that the question of definitions in mathematics assumed greater and greater importance for Peano, and he was never satisfied with the 'answers' given.

A final postcard from Peano, dated 7 January, 1903, shows that Frege had not forgotten him, for in it Peano thanks Frege for the gift of Vol. 2 of the *Grundgesetze*. Peano says he intends to study it, and rather brazenly asks Frege for a copy of Vol. 1, saying he had read it in the university library, which he had had buy it. (Now, he says, he must give Vol. 2 to the library.) Peano says he is sending a couple of things to Frege and urges him to translate his logical propositions into the symbolism of the *Formulaire*, i.e. into Peano's symbolism. (Peano remarks offhand that he, Peano, will not do it.) Peano promises to send the latest edition of the *Formulaire* as soon as it is ready. (This would be the 4th edition of 1902–03.) The tone is friendly, but shows Peano's complete confidence in the success of his own symbolism – and painfully little understanding of the impossible task he was asking of Frege.

What has been reported here is probably the extent of their contact. Frege rightly concluded in his 1896 letter that their aims in creating an ideography were distinct. Frege's goal was an investigation into the foundations of mathematics. Peano wished to create a symbolism that would be adequate for expressing mathematical theories. Peano recognized this distinction in what was essentially his last systematic formulation of mathematical logic, the 'Formulas of mathematical logic' [107] of 1900. He lists "several authors who have applied this ideography to different mathematical theories," and then adds: "Mr. Frege reached in his own way, in 1893, an ideography by which he has expressed propositions on the idea of number."

CHAPTER 11

PEANO ACQUIRES A PRINTING PRESS

In the three years between the First and Second International Congresses of Mathematicians, Peano was not only occupied with the continuing work on the *Formulario* and his usual teaching duties, but he also became involved in the activities of Mathesis, an organization of secondary school teachers of mathematics, which had been founded by Aurelio Lugli, Francesco Giudice, and Rodolfo Bettazzi in 1895. Both Peano and Padoa, who had graduated from the University of Turin in 1895 and was teaching in Pinerolo, 23 miles southwest of Turin, attended the meetings of the Turin section of Mathesis as early as 28 February, 1897. Burali-Forti joined the organization the following academic year, so that, with Giudice and Vacca in attendance, Peano and his followers were able to dominate the first congress of Mathesis, which took place in Turin in September 1898.

After the close of the First International Congress of Mathematicians in Zurich in August 1897, Peano returned to Turin as the acknowledged leader of the 'pasigraphic movement', as Schröder had called it. Giovanni Vacca was now his assistant at the university, but his primary role, in Peano's eyes, was probably as a collaborator on the *Formulario* project. Vacca remained Peano's assistant and collaborator for five years.

Other than a report (in collaboration with D'Ovidio and Segre) [95], Peano published only one article in the fall, 'General remarks on ordinary differential equations' [94]. In it he reviews the history of existence proofs for differential equations from the time of Cauchy, and asserts his priority over Emile Picard to the method of 'successive approximations', or 'successive integrations', as Peano preferred to call it. Peano had used this method for the first time in 1887 [10]. He remarks here that Picard first used it in 1891. (In fact, Picard had already given an example of successive approximations in 1888 and, at that time, gave credit for the method to H.A. Schwarz.) Picard seems to have taken no public notice of Peano's claim, which was repeated several times by him and his followers.

In January 1898, there appeared the second number of Vol. 6 of the *Rivista*, containing Frege's letter and Peano's reply [97], already discussed. There was also a note [96] giving tables of correspondence between the formulas of logic of the first two volumes of the *Formulaire*.

78

At the Academy of Sciences of Turin, on 13 March, Peano presented his 'Analysis of vector theory' [98]. The opening paragraphs give a description of the work:

To explain the theory of vectors, as well as the geometrical calculus, one ususally presupposes a greater or less knowledge of geometry.

I propose in this work to examine those ideas that are found in the theory of vectors, and to classify them into primitive, which are obtained from an observation of physical space, and into derivative, which are defined; and to examine which propositions one must assume as primitive and which can be deduced as consequences, using purely logical processes, without relying on intuition.

The result is a theory of vectors developed without presupposing any previous study of geometry, and seeing that the whole of geometry can be derived from the theory of vectors, there follows the theoretical possibility of substituting the theory of vectors for elementary geometry.

Frege had remarked: "The convenience of the typesetter is not the highest good." Peano, however, had shown his concern for the practical problems involved in printing his symbolism by insisting that the formulas be written on a single line. That he still met with difficulties is shown by his complaint over the delay in publishing Vol. 6, No. 2, of the *Rivista*. Whether because of this or for other reasons, he now became intimately involved with printing, by his purchase, in April 1898, of a printing press, which he installed in his villa in Cavoretto. As the press itself has a history that involves one of Peano's former professors, it seems worth the telling.

Francesco Faà di Bruno was a student in Paris of Cauchy, whom he admired, not only for his genius, but also for his religious fervor and his philanthropy. After returning to Turin, he founded several charitable institutions, including a home for poor girls, all of these being grouped together into a single organization called the *Conservatorio del Suffragio*. He wished to provide an occupation for these girls, and had the novel idea that they could become excellent typesetters. He therefore installed a printing press in the *Suffragio*, and a number of books were printed at the Tipografia Suffragio, including the second of a projected three-volume text on functions that he was preparing in his last years. (This volume, in French, treated elliptic functions. The other two volumes were never completed.) The printing press remained at the *Suffragio* until 1898, when it was purchased by Peano for 407 lire. Appropriately, one of the last works printed at the Tipografia Suffragio was a biography of Faà di Bruno, written by his friend and heir, Father Agostino Berteu.

Although the actual printing of the remaining editions of the *Formulario* was done in Turin, the type was set at Peano's villa in Cavoretto. The same was true of the volumes of the *Rivista*. A first result of this was the publication, with the date 9 August, 1898, of Part 2 of Volume 2 of the *Formulaire* [99]. This has the title 'Arithmetic', but it also included an extract from Part 1, so as to make it self-contained. With this work we now have the definitive statement of the five postulates, 1 being replaced by 0.

The first congress of secondary school mathematics teachers was organized by the Mathesis Society and took place in Turin from 9 to 14 September. Rodolfo Bettazzi, the president, was unable to attend. Giudice, in his name, proposed D'Ovidio for the president of the congress, and Peano and Segre for vice-presidents, as well as Castellano and Vacca for secretaries. All were elected unanimously, and as D'Ovidio was absent, Peano presided over the opening session. On 13 September, Peano held a 'Conversation on the *Formulario di Matematica*' at the university. At that conference he presented the recently published Part 2 of Vol. 2 of the *Formulaire*, the presentation being followed by "lively applause" (according to the official minutes of the congress.)

On Wednesday afternoon, 14 September, Peano again presided at the fifth and concluding session of the congress. Luigi Certo read a report on suggested changes to be made in the university program of future secondary teachers. Among other things, he insisted that a course in mathematical logic be instituted and required of all students of mathematics. The partisan secretaries recorded that "the reading of this report was received by very lively and prolonged applause." Peano then thanked the speaker, in the name of all, and said that the faith that Prof. Certo showed in the triumph of mathemaical logic was for him one of the greatest satisfactions. Thus the congress closed on a note of triumph for Peano and his followers, but how could it be otherwise, since there were present, besides those already mentioned, Padoa and the very vocal Burali-Forti?

The next number of the *Rivista* contained further additions and corrections to the *Formulaire* Vol. 2 [100] by Peano and others. Peano followed this with remarks 'On §2 of the Formulaire Vol. 2: Arithmetic' [101]. Here, his experience as a typesetter is alluded to. He says, in fact: "The notations of mathematical logic were created so as not to present typographical difficulties," and "These simplifications, by which formulas are composed the same as an ordinary text, will be much appreciated by whoever has a knowledge of the art of typography; some were already accepted, and others will be by whoever has to print at his own expense a book containing many formulas."

In this article, too, Peano mentions that he has begun using the *Formulario* in his university instruction "with excellent results," and he now thinks it possible to write an arithmetic and algebra text for use in the secondary schools, that would make consistent use of mathemtical logic. Such a book was actually published in 1902 [118], by which time he should have seen the disastrous effect on his teaching of his use of "the symbols," as his students named it, but he was ever the optimistic propagandist. To encourage the use in the schools of Vol. 2, Part 2, as a book of consultation — "like a table of logarithms" — he persuaded his publisher to sell it to teachers at cost.

Peano closes his article with an appeal to anyone interested in mathematics to collaborate on the *Formulario*. He indicates where improvements may be made and where entire theories are to be written; he gives suggestions on how to proceed, offering himself and the library of the *Rivista* for consultation; and he closes with a set of seven 'rules for the *Formulario*.' We mention two: No. 2 says that propositions will be published with the name of the person proposing them, "who then makes himself responsible for the exactness and importance of the propositions." No. 7 offers collaborators a complimentary subscription to the *Rivista* and the *Formulaire* then in preparation and, besides, "as many copies of the fascicle containing the additional propositions as there are propositions added." Clearly, this was not a money-making project.

At the 13 November session of the Academy of Sciences, Peano gave the Academy a copy of the German translation of his *Calcolo differenziale* of 1844 and presented for publication his 'Binary numeration applied to stenography' [103]. This curious note traces bits of the history of binary numeration and discusses some of its advantages (without, however, wanting to substitute it for the usual decimal system.) He then points out that the common prejudice against binary numeration, namely that it takes too much space to write, is false, since a single eight-pointed figure can have $2^8 = 256$ variations (omitting or including rays.) He then introduces various combinations to represent sounds of the Italian language and shows how these can be used for syllables, and he develops this into a complete system. An ordinary pen is inconvenient for writing this, he admits, but thinks that with a brush one could already write the binary system faster than ordinary writing. With a machine made for this purpose, real speed could be attained — and he has manufactured such a machine! One wonders how serious he was about this project. At any rate, there were too many other things to be concerned about. As editor of the *Formulario* project, he was "absorbed in various affairs," as he complained in his appeal for collaborators.

He referred to this article the following year, in a reply to a suggestion in the *Intermédiaire des Mathématiciens* for 'numerals' to be used in an arithmetic to base 20. Peano points out that combinations of base 2 numerals furnish the simplest means of representing numbers and anything numerable, "for example, the sounds of a language" [106.1].

The recently published Part 2 of the *Formulaire* contained the theory of rational numbers; irrationals would be discussed in Part 3. The debate on how best to introduce irrationals in the secondary instruction had already begun at the sessions of Mathesis. On 26 December it was discussed by Peano, Burali-Forti, and others; on 26 February, 1899, they returned to the topic, Burali-Forti suggesting the elimination of any treatment of prime numbers from the secondary schools, Peano wanting to keep it for its "logical beauty" (and he gave an example from a book that he "just happened to have at hand"!) On 28 March, they were at it again. Once again Peano showed himself the diplomatic peace-maker. He pointed out that each method discussed had its good side, and suggested that each prepare for a future session a complete theory of irrationals, according to his own view, and that from a confrontation of these theories would emerge a collective theory, in which not only rigor, which every scientific as well as didactic theory must have, but step by step with it would go simplicity.

Peano followed up his own suggestion by publishing in the 20 May issue of the *Rivista* an article on the theory of irrational numbers [105], in which he traces the history back to Plato and Euclid. His procedure is essentially that of Dedekind and of Peano's *Arithmetices principia*. There is an interesting discussion of whether or not real numbers are to be identified with segments of rational numbers. The distinction is discussed in terms of the difference between nominal definitions and definitions by abstraction. Peano choses the latter, as this was more in conformity with current usage and with Dedekind. He says: "In this *erschaffen* (create) is exactly indicated that the real number is considered as an entity different from the section, or segment." (Further concern for the classroom is shown in 1899 by a letter to the editor of a journal for teachers, in which he gives didactic reasons for defining 0^0 to be equal to 1 [104].)

During 1898, Peano's four notes to the *Intermédiaire des Mathématiciens* further reflect his preoccupation with the *Formulario*. In answer to a request for an elementary proof that $(1 + 1/m)^m < 3$ for all integral values of m [103.1], he remarks that he had given this in his calculus text of 1893, and he quotes from the *Formulaire*. He furnishes historical information on the introduction of fractional exponents [103.2]. He appeals to the readers of the *Intermédiaire* for help in the *Formulario* project, and offers to send page proofs of the part being printed to anyone interested [103.3]. To a reader who wants to know the best French and foreign tables of prime numbers, Peano gives bibliogrphic information on five such [103.4].

The Academy of Sciences of Turin met about eighteen times each academic year. From the year of his election, Peano had attended over half the sessions, but this spring (1899) his attendance began to fall off; he even missed the 12 March session, at which the Swedish mathematician Gösta Magnus Mittag-Leffler was present. (There was a low in attendance the following year, after which it increased again.) Although Peano did not attend the 14 May session, he sent gifts: a booklet by Domenico Tessari, the second part of the German translation of Peano's *Calcolo differenziale*, and the German and Polish translations of his 'Essay on geometrical calculus' [90].

Part 3 of Vol. 2 of the *Formulaire* [106] appeared in the summer. In addition to the new material, Part 3 also reproduces, with corrections and additions, Part 2 published the previous year, but contains only enough of Part 1 to make it readable. In this work the decimal ordering of the propositions, begun in Part 2, is systematically carried out "with the purpose of making interpolation easier." The mathematical content is greater than that of Vol. 1, although the chapters on measure and algebraic numbers, due to Burali-Forti and Gino Fano, respectively, have been dropped, even though Peano says that "to complete the 'mathematical encyclopedia', it is necessary to add" those theories.

In the preface, Peano gives credit to Burali-Forti, Castellano, Chini, Padoa, Vacca, Vailati, and Vivanti for the work on the previous editions. He then says: "In the theory of limits and series, we have profited by the work of R. Bettazzi and F. Giudice, published in Vol. 1, sections 7 and 8. The other sections are in general here reduced to symbols for the first time". Ugo Cassina sees in this an explicit declaration that Peano assumed personal responsibility for this new edition of the *Formulaire*, although taking into account the collaborators' suggestions, which for that matter were limited to correcting typographical errors or adding an occasional new proposition. Padoa and Vacca are exceptions; their contributions became numerous enough to be given special symbols. To Vacca is also due the citation of Leibniz' manuscripts on mathematical logic, which he inspected in Hannover, and the general bibliography. Indeed, Vacca "powerfully aided me in the whole work."

CHAPTER 12

THE SCHOOL OF PEANO

In July 1899, Tommaso Boggio graduated 'with high honors' in pure mathematics and in the fall he became university assistant for projective and descriptive geometry. He was another addition to the 'school' of Peano. In 1932, Ugo Cassina gave a list of 45 Italians who belonged to the school of Peano. (This list is given in the appendix.) In this chapter the careers of seven of Peano's early followers are described briefly. At the turn of the century these seven were perhaps the most committed to Peano's programs and most of them have already been mentioned several times. They are: Vailati, Castellano, Burali-Forti, Padoa, Vacca, Pieri, and Boggio.

Giovanni Vailati was born in Crema (Cremona) on 24 April, 1863, and he died in Rome on 14 May, 1909. After attending boarding schools in Monza and Lodi he enrolled in the University of Turin in 1880, graduating in engineering in 1884 and in mathematics in 1886. Then followed a period of independent study, especially of languages (his writings show a proficiency in Greek, Latin, English, French, German, and Spanish.) He was Peano's assistant from 1892 to 1895, when he became an assistant in projective geometry and later honorary assistant to Vito Volterra. In 1899 he requested a secondary school appointment and was at first sent to Syracuse, transferring to Bari in 1900, to Como in 1901, and to Florence in 1904. Vailati came of a Catholic family, but lost his faith during his early university years. Throughout his life he had affectionate and devoted friends. He never married. His premature death was attributed to heart trouble, complicated by pulmonitis.

Vailati's first ten publications, dealing principally with mathematical logic, were published in the *Rivista de Matematica*, but he gained international recognition with the publication of three essays in the history and methodology of science, originally given in the years 1896—1898 as introductory lectures to his course at the University of Turin in the history of mechanics.

Vailati received most attention during his lifetime as the leading Italian exponent of pragmatism. His philosophical position was closer to that of C.S. Peirce than to the more popular William James, but it remained distinct, individual, and original.

On 17 May, 1909, Peano published an obituary of Vailati in the newspaper

Gazzetta del Popolo (of Turin), which was later reprinted in the *Bollettino di Matematica* [142.1]. In it he recalled especially Vailati's work on the commission for school reform, noting that Vailati had traveled to several countries to study their school systems. Finally, "he was modest, friendly with all, and universally esteemed, both for his learning as also for his personal qualities."

Peano wrote perhaps the only obituary of *Filiberto Castellano* [190]. He tells us that Castellano came of "humble origins" in Pietra Marazzi (Alessandria) and received his degree in mathematics at the University of Turin in 1881. He was at first assistant in algebra and analytic geometry, and then (Peano's) assistant in calculus, until he was named professor at the Military Academy (presumably in 1892, when Vailati became Peano's assistant.) His lessons on rational mechanics at the Military Academy were published in 1894. Peano says this was the first such text to make systematic use of vectors, noting that at first the vectorial method met with the opposition of the conservatives, but that Castellano lived to see its triumph, being adopted in almost all the Italian universities. Castellano was one of the founders of the *Rivista di Matematica* and was on the administrative committee of Mathesis for the two-year periods 1898–1900 and 1902–04.

Castellano, like Peano, had a villa in Cavoretto. He died on 24 January, 1919, at the age of 58. (Thus, he must have been almost exactly Peano's age.)

Cesare Burali-Forti was born in Arezzo on 13 August, 1861, and died in Turin on 21 January, 1931. He graduated from the University of Pisa in December 1884. After teaching briefly in a secondary school, he transferred to Turin, where he taught analytical projective geometry at the Military Academy from 1887 until his death in 1931. He early attempted to obtain the *libera docenza*, but failed, despite the fact that Peano was one of the judging committee, because of his insistence on vectorial methods. Peano, of course, defended him, but was unable to persuade a majority. He never again tried to obtain it and so never held a permanent university position. He was, however, assistant to Peano during the years 1894–96.

Already under the influence of Peano, he gave in 1893–94 an informal series of lectures at the university on mathematical logic, which he published in 1894. Many of his publications were highly polemical, but in his family circle and among friends he was kind and gentle. He loved music, Bach and Beethoven being his favorite composers. He was a member of no academy. Always an independent thinker, he asked that he not be given a religious funeral.

The name Burali-Forti has remained famous for the antinomy he discovered in 1897 in his critique of Georg Cantor's theory of transfinite ordinal numbers. This result went almost unnoticed until Bertrand Russell published a similar antinomy in 1903. It should be noted, however, that Cantor was already aware of these difficulties in 1895 and had written of them to Hilbert in 1896.

Burali-Forti's most valuable mathematical contributions were his studies devoted to the foundations of vector analysis, linear transformations, and their various applications, especially in differential geometry. As has been indicated earlier, he contributed extensively to the *Formulario* project. A long collaboration with Roberto Marcolongo — their friends called them "the vectorial binomial" — was in part broken by their divergence of view over the theory of relativity, the importance of which Burali-Forti never understood. Besides his scientific publications, he wrote many school texts. In all, his publications total more than 200.

Alessandro Padoa was born in Venice on 14 October, 1868, and died in Genoa on 25 November, 1937. He attended secondary school in Venice, engineering school in Padua, and the University of Turin, from which he received a degree in mathematics in 1895. Although he was never a student of Peano, he was an ardent disciple and, from 1896 on, a collaborator and friend. He taught in secondary schools in Pinerolo, Rome, Cagliari, and (from 1909) at the Technical Institute in Genoa. He also held positions at the Normal School in Aquila and the Naval School in Genoa, and, beginning in 1898, he gave a series of lectures at the Universities of Brussels, Pavia, Berne, Padua, Cagliari, and Geneva. He gave papers at congresses of philosophy and mathematics in Paris, Cambridge, Livorno, Parma, Padua, and Bologna. In 1934 he was awarded the ministerial prize in mathematics by the *Accademia dei Lincei* (Rome).

Padoa was the first to give a method for proving that a primitive term of a theory cannot be defined, in the system, by the remaining primitive terms. This was presented in his lectures in Rome in early 1900 and made public at the International Congress of Philosophy in Paris later that year. Although Padoa was Jewish, since he died in 1937, on the eve of the racial persecutions, he was not disturbed by them.

Giovanni Enrico Eugenio Vacca was born in Genoa on 18 Novermber, 1872, and died in Rome on 6 January, 1953. He graduated in mathematics from the University of Genoa in 1897. Even before beginning university studies, he was active politically and assisted in the founding of the Italian Socialist Party in Genoa in 1892. (He was a friend, among others, of Filippo Turati.) In the period of reaction that followed, he was sentenced to one year's banishment from Genoa in 1897. Ugo Cassina believed that Peano met Vacca at the Zurich congress in August of that year, but it seems unlikely that Peano would not have selected his university assistant for the coming academic year by then. Although Cassina says that Vacca became Peano's assistant in 1898, the university yearbook already lists him as having this position in 1897. During the summer of 1899, Vacca went to Hannover to study the unpublished manuscripts of Leibniz. His note on this matter prompted Louis Couturat, whom he met in August 1900 at the International Congress of Philosophy, to prepare for publication in 1903 a volume of short works and fragments from the previously unpublished manuscripts of Leibniz. Vacca remained Peano's assistant until 1902, when he returned to Genoa.

In the period 1902–04, he was a member of the municipal socialist council and the national party administration. He was also an assistant at the university and gave a course in mathematical logic there. He returned to Turin in 1904 for one more year as Peano's assistant. (He had actually been listed the previous year as an honorary assistant.)

By this time his interest in Chinese became so strong that he moved to Florence to study that language under Carlo Puini, professor of the history and geography of eastern Asia. His interest really began in 1898, during the exposition in Turin, where, at an exhibit of sacred art, he met two religious missionaries to China, who showed him his first Chinese books and gave him several lessons. It is also easy to believe that he was attracted by the conceptual similarity between Chinese writing and Peano's ideography. It was during this period in Florence that he became good friends with Vailati, whom he had already met in Turin. In a postcard to Peano, dated 11 April, 1906, he has a remark about the history of a proposition of logic, and adds: "Greeting from Vailati, who is here with me, and from your affectionate Giovanni Vacca."

During the years 1907–08, he made an extended trip to western China, spending an entire year in the city of Cheng-tu. Back in Florence, he received the *libera docenza* in Chinese language and literature in 1910. From 1911 to 1922 he taught Chinese literature at the University of Rome. In 1922 he succeeded Puini in Florence, as professor of the history and geography of eastern Asia, and in 1924 he transferred to the University of Rome as professor of the same subject, where he also taught Chinese language and literature until his retirement in 1947.

In 1921, Vacca married Virginia De Bosis, an expert in Arabic studies; they had two children.

Vacca's interest in mathematics continued side by side with his interest in things Chinese. Of some 130 publications, 47 treat of the Orient, 38 are in the mathematical sciences, and 45 deal with the history of mathematics and science.

Mario Pieri was born in Lucca on 22 June, 1860, and died in S. Andrea di Compito (Lucca) on 1 March, 1913. He began his university studies in 1880 in Bologna, where Salvatore Pincherle was among the first to recognize his talent, but having obtained a scholarship at the Scuola Normale Superiore in Pisa in November 1881, he completed his university studies there, receiving his degree on 27 June, 1884. He taught briefly at the technical school in Pisa and then went to Turin, where he became professor of projective geometry at the Military Academy and also, in 1888, assistant in projective geometry at the University of Turin, holding both positions until 1900. He received his *libera docenza* from the university in 1891 and for several years taught an elective course in projective geometry there.

From 1900 he taught projective and descriptive geometry at the University of Catania, transferring in 1908 to Parma, where in the winter of 1911 he began to complain of fatigue, his fatal illness (cancer) being diagnosed a few months later.

For ten years following his first publication in 1884 Pieri worked primarily with projective geometry. From 1895 on he studied the foundations of mathematics, especially the axiomatic treatment of geometry. Pieri had made a thorough study of Christian von Staudt's geometry of position, but he was also influenced by his colleagues at the Military Academy and the university, Peano and Burali-Forti. He learned symbolic logic from the latter, and Peano's axiom systems for arithmetic and ordinary geometry furnished models for Pieri's axiomatic study of projective geometry.

In 1895, Pieri constructed ordinary projective geometry on three undefined terms: *point, line*, and *segment.* In their axiom system for ordinary geometry Pasch had used four undefined terms, and Peano three. Pieri reduced this number to two: *point* and *motion.* In 1905, he gave the first axiom system for complex projective geometry which is not constructed on real projective geometry.

In a note of 1907, 'Sopra gli assiomi aritmetici,' he selected as primitive notions *number* and *successor* of a number, and characterized them with a system of axioms which from a logical point of view simplifies Peano's theory.

In 1911 Pieri may have been on the point of beginning a new phase of his scientific activity. He was then attracted by the vectorial calculus of Burali-Forti and Roberto Marcolongo, but he left only three notes on this subject.

Pieri became one of the strongest admirers of symbolic logic and although his works are mostly published in more ordinary mathematical language, we know from the statements of colleagues and from his own statements that he considered the use of Peano's symbolism of the greatest help not only in obtaining rigor, but in actually deriving new results.

Tommaso Boggio was born in Valperga Canavese (about 26 miles north of Turin) on 22 December, 1877, and died in Turin on 25 May, 1963. In October 1895 he won, over twelve others, a competition for a post in the Collegio delle Provincie. (Peano was one of the judges in the competition.) In July 1899, as we have already noted, he graduated 'with high honors' in pure mathematics, and in the fall he became assistant for projective and descriptive geometry, and immediately after *incaricato* for a year. In 1903, he received the *libera docenza* in mathematical physics and taught various courses until 1905, when he became Extraordinary Professor of the Mathematics of Finance in Genoa. In 1908 he became Extraordinary Professor of Rational Mechanics in Messina, where he fortunately escaped the disastrous earthquake of 28 December that same year (in which approximately 78,000 people died.) The university being destroyed in the earthquake, he was called to Florence and then Turin, where from November 1909 he was Extraordinary Professor of Higher Mechanics, and Ordinary from 1911 until 1942, when he transferred to Complementary Mathematics until his retirement in 1948. In addition to his professorship, he also taught many courses at the Military Academy and he gave private lessons, even to his own university students, a fact which lowered him in the estimation of many.

Boggio suffered many family difficulties. His wife is said to have been of little support to him, a daughter died during World War II and his second son died at the age of 46 (the first son had emigrated to Argentina), leaving him to care for his daughter-in-law and two grandchildren.

Boggio's scientific activity was, from around 1910, mainly concerned with classical mathematical physics, especially potential theory.

Of the seven men whose biographies have just been sketched, all but Padoa served as university assistants to Peano. The work of all of them was directly inspired by Peano and all preserved their admiration and affection for him,

Vacca dedicating one of his last works to his three favorite professors: Peano, Puini, and Vito Volterra. All of them made valuable contributions to the fields of interest to Peano: logic, the foundations of mathematics, vectors, although Castellano's contribution was not as notable as that of the others. These seven then, are representative of a real school — the school of Peano. Having named them, one immediately wants to add other names, but we must move along in our story, and at any rate we shall meet some of them again shortly in Paris.

CHAPTER 13

PARIS, 1900

Before the beginning of the fall term in 1899, Peano was on holiday in Naples from 20 September to 10 October. During the academic year 1899– 1900, he attended only three sessions of the Academy of Sciences of Turin; we may presume that he continued to be occupied with the *Formulario* project and, no doubt, with preparations for the summer congresses in Paris. In June 1900, classes at the university were suspended for a few days at the beginning of the month because of student agitation to have exams postponed so that they could go home to vote. This right was then granted to those who could show that they had, in fact, voted, but only four students took advantage of it.

The Italian pavilion at the International Exposition in Paris was opened on 2 May, 1900, but the interest of Peano and his followers centered on the international congresses of mathematics and philosophy, to be held in August – especially the International Congress of Philosophy. Of this congress Hans Freudenthal has remarked: "In the field of the philosophy of sciences the Italian phalanx was supreme: Peano, Burali-Forti, Padoa, Pieri absolutely dominated the discussion." It may be added that this 'phalanx' was backed up by Vailati and Vacca, for in addition to the first four, Vailati also read a paper. Concerning Vacca, Louis Couturat wrote (it is the opening of the preface of his *Opuscules et fragments inédits de Leibniz*):

Our work on *The Logic of Leibniz* was almost completed (at least we thought so) when we had the pleasure, at the International Congress of Philosophy (August 1900), of making the acquaintance of Mr. Giovanni Vacca, at that time mathematical assistant at the University of Turin, who had examined, the year before, the manuscripts of Leibniz preserved in Hannover, and had extracted from them several formulas of logic inserted in the *Formulaire de Mathématiques* of Mr. Peano. It was he who revealed to us the importance of the unpublished works of Leibniz, and inspired us with the desire to consult them in turn.

If meeting the Italian Vailati had an effect on Couturat, meeting Peano had an even greater effect on Bertrand Russell, who wrote in his *Autobiography* [1967, 217–218]:

The Congress was a turning point in my intellectual life, because I there met Peano.

91

I already knew him by name and had seen some of his work, but had not taken the trouble to master his notation. In discussions at the Congress I observed that he was always more precise than anyone else, and that he invariably got the better of any argument upon which he embarked. As the days went by, I decided that this must be owing to his mathematical logic. I therefore got him to give me all his works, and as soon as the Congress was over I retired to Fernhurst to study quietly every word written by him and his disciples. It became clear to me that his notation afforded an instrument of logical analysis such as I had been seeking for years, and that by studying him I was acquiring a new and powerful technique for the work that I had long wanted to do. By the end of August I had become completely familiar with all the work of his school. I spent September in extending his methods to the logic of relations. It seems to me in retrospect that, throughout that month, every day was warm and sunny.

Russell wrote up his ideas and sent the manuscript of 'Sur la logique des relations avec des applications à la théorie des séries' to Peano that fall. Peano delayed his reply, but finally wrote on 19 March, 1901: "I shall publish directly your interesting memoir, which fills a gap between the work of Peirce and Schröder on the one hand and the *Formulaire* on the other. Let me congratulate you on the facility and precision with which you manage the symbols of logic." The article was published in the *Rivista di Matematica* in July 1901 and is the only article in that volume (1900–01) not directly connected with the *Formulario* project. The *Rivista di Matematica* published a second article by Russell ('Théorie générale des séries bien ordonées') in 1902. Russell also contributed to the journal financially; for example, in 1905 when the subscription price was 11 francs, he sent 128 francs.

In a less dramatic sense, the congress was also a turning point for Peano, for one of the works which Peano would surely have given Russell was his latest, *and last* systematic development of his mathematical logic [107]. It occupied the first 41 pages of Volume 7 of the *Rivista* and carried the date 20 July, 1900. The following year it would be reproduced almost without change in Volume 3 of the *Formulaire*. This same issue of the *Rivista* also contained 'Additions to the *Formulaire*', four pages of which are by Peano [108].

If the origin of the First International Congress of Mathematicians is obscure, it is quite clear that the initiative for the First International Congress of Philosophy came from the journal *Revue de Métaphysique et de Morale*. The 'patronage committee' of the congress included six Americans (F. Adler, J.M. Baldwin, P. Carus, J.E. Creighton, J. Royce, and J.G. Schurman) and six Italians. Peano and Vailati were both on the committee. Among the Germans were G. Cantor, M. Cantor, R. Dedekind, G. Frege, F. Klein, and E. Schröder.

The congress opened on Wednesday, 1 August, and already at the first session, which was held jointly with the congress for secondary education (concerned especially with the teaching of philosophy) Padoa joined in the discussions. He defended the teaching of the history of philosophy, considering this as a defense against dogmatism in philosophy. The next morning, at the session of Section 3: Logic and history of the sciences, Peano showed his love of homely examples by recalling (this was after a talk on physical determinism and human liberty) the attempts made to reconcile determinism and contingency in the domain of science, and especially that of Boussinesq, based on the existence of singular solutions in certain problems of mechanics, an example of which is the following: Let a razor blade be horizontal and turned upward, and let a material point be placed in equilibrium on it. On which side will it fall?

In the afternoon there was a general session for logic and history of the sciences. In the discussion following a paper of Poincaré, chairman of the session, Padoa revealed his interest in the foundations of mathematics. The distinction between axioms and definitions, he said, has only a logical and subjective value; in the real world there are only facts, all presented on the same level. He stressed the reciprocal independence between the logical and the real, the complete split between the two.

Peano read his paper on 'Mathematical Definitions' [109] on the morning of Friday, 3 August, at the session of Section 3 that was presided over by Jules Tannery. (This paper received a Polish translation in 1902 [109'].) He had already discussed the question of definitions in mathematics in his writings on mathematical logic, but this was his first systematic treatment of the subject. There would be others. Indeed, this was the one 'philosophical' question that continued to interest him until the end. His last systematic work on this subject had the same title and appeared in 1921 [193], but he was never satisfied that the question was settled, so in 1931, only shortly before his death, he assigned this to L. Geymonat as a minor thesis topic for a degree in mathematics.

For Peano, definitions are basically equalities that satisfy certain 'laws', one of which is the 'law of homogeneity', which he explains as follows:

A proposition must first of all be complete, intelligible by itself, apart from the other propositions of the text. I understand a definition in this way, in contrast to some authors, who call *definition* the most important part, or the formula, contained in the complete proposition.

Next, the two members of the equality making up a definition must contain the same real variable letters.

Thus the formula: $0 = a - a$ is not a definition, since it is not a complete proposition; it has not been said what value we give to the letter a.

The proposition:

Let a be a number; we have: $0 = a - a$

is a complete proposition; but it may not be taken as a definition of the sign 0, since it is not homogeneous. In fact, the first member is a constant symbol 0, the second is a function of the variable letter a.

The proposition: $0 =$ (the constant value of the expression $a - a$, whatever the number a is) is a homogeneous equality, since even though the letter a appears in the second member, it is only an apparent [variable], since the value of the second member is not a function of a.

After the talk, E. Schröder commented that one should not impose on definitions conditions that are too restrictive. For example, the *logical zero* or *néant* could be defined by the formula: $0 = aa_{\prime}$ [where a, is the negation of a] whatever a may be. J. Tannery had objections to any definition of the *arithmetic zero*. Peano insisted on the necessity of homogeneity and was supported in this by Padoa.

Bertrand Russell, commenting on a definition of Schröder, wrote in a note to Norbert Wiener: "There is a need of a notation for 'the' . . . . It was a discussion on this very point between Schröder and Peano in 1900 at Paris that first led me to think Peano superior" [quoted in Grattan-Guinness 1975, 110].

Typically, Peano draws practical consequences from his theory. The closing lines from this brief note are worth quoting.

Finally, we remark that every sign or word defined in a science may be suppressed, on the condition of replacing it by its value. In other words, every definition expresses an abbreviation, which is theoretically unnecessary; in practice it may be convenient and even indispensable to the progress of the science.

The suppression in a theory of a defined sign is a very useful exercise for recognizing the exactness of the definition, for if one cannot replace it by its value everywhere, the definition is incomplete. This serves also for judging the usefulness of the definition, for if the propositions in which the sign occurs are not made too long when they are suppressed, it is suitable to suppress them. In treatises of mathematics there are a good many words that it would be suitable to suppress.

Although it appears that not all of them were present, this day was definitely carried by the Italian phalanx. Peano's paper was followed by Burali-Forti's paper on 'The different logical methods of defining real numbers,' which was summarized by Couturat. Padoa was next, and then Couturat summarized Pieri's paper on 'Geometry considered as a purely logical system.'

Padoa's paper was entitled 'Essay of an algebraic theory of whole numbers, preceded by a logical introduction to any deductive theory,' but his talk was mainly about the 'Introduction'. In it he discussed two parallel problems concerning a deductive system: (1) Are the axioms independent? Or can one axiom be derived from the others? (2) Are the primitive terms irreducible? Or can one be defined by means of the others? The first of these had, as we have seen, already been discussed by Peano and developed by him into a standard method for proving the independence of a set of postulates. Padoa shows here how to answer the second question, with a technique that has come to be called Padoa's Method. He defines a system of undefined symbols to be *irreducible* with respect to the system of axioms if it is impossible to give a symbolic definition of any of the undefined symbols. To prove this irreducibility, it suffices to find an interpretation of the system of undefined symbols, such that they continue to satisfy the system of axioms when the meaning of only one of them is changed; and this for each of the undefined symbols, taken separately.

Padoa's paper is dated: Rome, 1 July, 1900, but he had already presented his method in lectures in Rome earlier that year. Although it took the development of model theory, much later, to bring out the importance of the method in the theory of definitions, Padoa was already convinced of its significance. He says, in fact, in his paper: "We can now settle completely (and, we believe, for the first time) a question of the greatest logical importance." (A proof of Padoa's method was given by Tarski in 1926.)

Couturat closed the session by recalling that Peano was the first to determine (in 1888) the number of propositions that can be stated concerning n simple classes. That was at the beginning of his interest in mathematical logic, but his interest had gone far beyond that, and it was his later development of mathematical logic that interested Bertrand Russell. Perhaps it was after this session that Russell introduced himself, or perhaps it was the next day, when Russell presented a paper that Hans Freudenthal says "bordered on a practical joke [Bierulk]." At any rate, Russell was impressed. He later wrote [1959, 65]:

It was at the International Congress of Philosophy in Paris in the year 1900 that I became aware of the importance of logical reform for the philosophy of mathematics. It was through hearing discussions between Peano of Turin and the other assembled philosophers that I became aware of this. I had not previously known his work, but I was impressed by the fact that, in every discussion, he showed more precision and more logical rigour than was shown by anybody else. I went to him and said, "I wish to read all your works. Have you got copies with you?" He had, and I immediately read them all. It was they that gave the impetus to my own views on the principles of mathematics.

Peano must have made a number of other acquaintances as well. Courturat, of course, had been in contact with him, but they may have met personally for the first time. It is probable that he also met Samuel Dickstein, who had founded the journal *Wiadomości Matematyczne* in Warsaw three years earlier.

The next day, at the session of Section 3, Peano again takes an active part in the discussion. Padoa remarks that this section "has perhaps been – and in any case numerically – the most important of the congress."

In the afternoon general session, Couturat brings up the question of an international language and pleads for its acceptance. (He does not indicate his choice, but he was known to be an Esperantist.) He was unanimously elected to represent the congress in the committee being organized by his friend Leopold Leau to study this question. Peano was no doubt in sympathy with Couturat's aim, but it is doubtful whether he was already actively interested in an international language. In any case, as we shall see, his choice would not be Esperanto.

The congress closed the following day, Sunday, 5 August, with the selection of delegates to make decisions about the next congress. The delegates nominated from the United States were: J.M. Baldwin, P. Carus, Mrs. Christine Ladd Franklin, Alexander Macfarlane, and J.G. Schurman. The delegates from Italy were: Carlo Cantoni, Mario Calderoni, Peano, and Vailati.

The Second International Congress of Mathematicians would open the next morning and delegates were already arriving for it. Like Peano, a number of mathematicians had attended the philosophy congress and simply stayed on. There was an informal reception for them that Sunday evening at the Café Voltaire, at which about half the delegates were present. (About 1000 mathematicians had been expected to attend the congress, but, whether because of the crowds that were supposed to be visiting the Exhibition and the rumored difficulty in obtaining accommodations or the July heat wave, only some 250 actually attended.)

The feeling of euphoria that closed the Zurich congress did not attend the opening of the Paris congress. The arrangements excited a good deal of criticism. Miss Scott, in her report to the American Mathematical Society, said that "members arriving in Paris had very great difficulty, during the first two days, in obtaining the necessary information in time for it to be of any service. The want of a common assembly room, where members might conveniently meet one another with or without concerted arrangement, was seriously felt. The arrangements in Zurich were so admirably complete in every point, that these defects were even more conspicuous by comparison." An Italian observer reported: "The congress was supposed to have opened at 2:00 p.m. on 6 August, but, by a certain hint of disorder that uniformly reigned and which contributed to making it more original, it was opened at 9:00 in the morning." Hilbert's talk had been scheduled for the opening session, but as he showed up late, it was postponed until the joint session of the morning of Wednesday 8 August.

Hilbert's paper proposed 23 problems which he believed would be, or should be, occupying the attention of mathematicians in the 20th century. In his talk at the congress he presented only ten of these, including No. 2, which asked whether it can be proved that the axioms of arithmetic are consistent. He was probably surprised when Peano announced, after this talk, that this question had already been answered (but there is no evidence that he made any effort to discover whether Peano was correct.) As reported by Miss Scott: "A more precise objection was taken to Mr. Hilbert's remarks on the axioms of arithmetic by Mr. Peano, who claimed that such a system as that specified as desirable has already been established by his compatriots Messrs. Burali-Forti, Padoa, Pieri, in memoirs referred to on pp. 3–5 of No. 1, Vol. 7 of the *Rivista di Matematica*." The author of the proceedings of the congress wrote: "Mr. Peano declared that the later Communication of Mr. Padoa will answer problem No. 2 of Mr. Hilbert." He was referring to Padoa's paper on 'A new irreducible system of postulates for algebra,' which was scheduled to be presented the following day at a session over which Hilbert presided. Unfortunately (for we might otherwise have learned Hilbert's opinion), another paper of Padoa ('A new system of definitions for Euclidean geometry') was scheduled at the same time in another section, and he chose to present the latter. He presented the 'postulates' paper the next day, but Hilbert did not attend that session.

Padoa's paper began by repeating the 'Logical introduction to any deductive theory' that he had presented the week before at the congress of philosophy. This time he merely summarized that section. Miss Scott found this one of the more interesting papers of the congress, although she seems to have felt that he spent too much time on the independence of the postulates. She wrote: "Mr. Padoa did not get beyond this definition, possibly because he had entered so minutely into the details of the proof of the independence of the seven postulates that he had exhausted his allowance of time."

George Bruce Halsted was also impressed with Padoa, listing him, in his report in the *American Mathematical Monthly*, as one of fourteen "among the most interesting personalities present." Peano did not present a paper at this congress, although the Italian reporter mentioned earlier (apparently

himself tinged with the disorder he reported) wrote that Peano presented a
paper on mathematical logic.

At this session Leopold Leau proposed the following resolution, the discussion of which, because of the importance of the subject, was postponed
for two days: (1) There is reason to adopt a universal scientific and commercial language. (2) The official academies are respectfully invited to work
together for the realization of this project. He further proposed the additional
resolution: The Congress agrees to name five members to the commission
being formed: (1) To ask the official academies to be willing to adopt a
universal auxiliary language; (2) To make the choice itself or by a committee
created by it, in case of a refusal by the academies.

On Friday, Laisant read a note of Charles Méray 'On the international
auxiliary language of Dr. Zamenhof, known under the name of 'Esperanto'.'
This led to a discussion of Leau's resolution. Those in favor of it included
Couturat, Laisant, Leau, Giovanni Boccardi, and Padoa. Those opposing the
resolution were led by Schröder, A. Vasil'ev, and Gian Antonio Maggi. Maggi's
argument was that mathematics already had a universal language, the language
of formulas, but he was an exceptional linguist, with a knowledge of a number
of ancient and modern languages, and so probably did not appreciate the
difficulties of the ordinary mathematician. Vasil'ev did appreciate the difficulties caused by the proliferation in the number of national languages being
used for mathematics, and in the end it was his resolution that passed, namely
"that the academies and learned societies of all countries study the proper
means for remedying the evils arising from the increasing variety of languages
used in the scientific literature."

Miss Scott assumed this meant that only English, French, German, and
Italian were to be used. She does not tell us how she voted, although she gives
some evidence of thinking the idea of an artificial language impractical. With
regard to systems of classifying the mathematical sciences, however, she thinks
the congress ought to pronounce in favor of some one, noting that: "Multiplication of systems of classification, like the multiplication of universal
languages, practically destroys the good of any and all."

The concluding banquet was held at noon on Sunday, 12 August, at the
Salle de l'Athenée-Saint-Germain, where Darboux was toastmaster, in the
absence of Poincaré, and toasts were proposed (according to Miss Scott) "to
those present, to the hosts, to the absent, to Mr. Darboux, and to the next
Congress." The next congress was to be held in Germany and, although Miss
Scott thought the banquet "very successful", no doubt there were those who
anticipated German organizational efficiency at the next congress.

A.N. Whitehead attended this congress, but Russell, having persuaded Peano to give him "all his works" at the congress of philosophy, left Paris as soon as that congress was over and returned home "to study quietly every word written by him and his disciples." Peano, too, now returned home and resumed work on the third edition of the *Formulaire*. In September, Russell extended Peano's methods to the logic of relations, and he sent Peano a long article on this for the *Rivista*. Peano put it in the next issue, but that did not appear until 15 July, 1901. The *Formulario* project of the *Rivista* had now taken priority. It was a case of the tail wagging the dog!

CHAPTER 14

THE DECLINE BEGINS

The third edition of the *Formulaire de Mathématiques* [110] is dated 1 January, 1901. It essentially marks the end of Peano's original researches in mathematical logic. The section on logic is almost verbatim the article on logic published the previous July in the *Rivista*, which was a new elaboration, and hence this edition differed greatly from the second edition of the *Formulaire*, at least in its treatment of logic. Its mathematical content differed little. Couturat gave it an ample review (nineteen pages) in the *Bulletin des Sciences Mathématiques* [(2) 25–I (1901), 141–159].

Peano gave a copy of the new *Formulaire* (which, although the title page says it was published in Paris, was, as usual, set with the type of the *Rivista* and printed in Turin) to the Academy of Sciences on 13 January. Then in February, taking advantage of the holiday for carnival, he made a trip to Rome and Tunis. He probably felt that this was a well-deserved vacation. He was at the height of his fame as mathematician and logician, and the Italian government recognized him on 25 April by making him a knight of the Order of Saints Maurizio and Lazzaro. (This order had been instituted in 1572 by Emmanuele Filiberto of Savoy, by uniting two ancient orders, for the purpose of protecting the faith from infidels and the sea from pirates. It was revived on 20 February, 1868, with changed rules, as a means of recognizing both civilian and military service.) Closer to home, however, all was not well, and on 26 May, 1901, Peano terminated his service to the Military Academy, having taught there for fifteen years. The termination was not voluntary.

The trouble was with his teaching. In the beginning, he was very good — this is shown by the excellent calculus text of 1893 [60] for the Military Academy — but as the *Formulario* project matured, it intruded more and more into his classroom teaching, so that he reduced the time given to the ordinary parts of calculus in order to devote more time to what the students called "the symbols." Costantino Botto, who is said to have been afflicted with 'Peanitis' because he admired Peano so much, was an engineering student in 1904–05, studying calculus with Peano. He later wrote [1934, 19]:

The textbook that Peano followed had become instead the *Formulaire*, from which he taught, with the greatest love and much patience, the first pages dedicated to the symbols of logic and then several lines of several other pages, dedicated to accurate definitions of the concepts, to the various operations, and several passages of various parts of mathematics. Only in the last months of the academic year did Peano arrive at briefly treating, and always with his symbols, the calculus by the system of vectors, and of explaining some applications of curves, with calculations of lengths, areas, etc. . . .

But we students knew that this instruction was above our heads. We understood that such a subtle analysis of concepts, such a minute criticism of the definitions used by other authors, was not adapted for beginners, and especially was not useful for engineering students. We disliked having to give time and effort to the 'symbols' that in later years we might never use.

Not only the students, but Peano's colleagues also objected to this method of teaching, and efforts were made to persuade him to change – to no avail. It was for this reason that he was relieved from his teaching position at the Military Academy in 1901. (He was replaced by Rodolfo Bettazzi, according to Botto. F.G. Tricomi thinks it was by Tommaso Boggio.) There were at the university those who would have liked to have Peano replaced there, but that was impossible. Finally, however, in 1907 the engineering school hit upon the idea of separating their students from students of the sciences, with whom they had previously shared the first two years of mathematics, so as to effectively remove them from Peano's teaching. Peano was thus replaced by Guido Fubini for the larger group of his students. (There were about 100 students in Botto's class.)

The statement of Botto is confirmed by Alessandro Terracini in his autobiography, and is the tradition F.G. Tricomi reports finding on his arrival at the University of Turin in the fall of 1925. In extraordinary contrast to this, however, is the report of Giovanni Vacca, read as part of a commemoration of Peano at a session of the Italian Society for the Progress of Science in 1932. He said:

His lessons, which varied every year, represented a continual effort to attain a more lucid exposition. I remember the first part of the course of 1903, which began with the method of the geometry of indivisibles of Bonaventura Cavalieri. I remember the lessons on the theory of irrational number, illustrated with Book V of Euclid, the lessons on the rectification of curves, starting from the exposition of Archimedes. Finally, I remember the reading of pages of Galileo and Torricelli on falling bodies, and the lessons on the calculus of variations, which interpreted in a new form the classical memoirs of Euler and Lagrange.

It should be remembered, however, that Vacca had been Peano's assistant for the years 1897–1902 and was an active collaborator on the *Formulario* project. (In 1903 he was 'honorary assistant'.) Furthermore, he shared Peano's interests in the history of mathematics. In short, there probably were always a few who could appreciate and share Peano's interests, but for the majority of students beginning the study of mathematics his classes were not helpful. In the light of this, one wonders why he was put in charge of the course in higher analysis in 1908, only to have it taken away from him two years later because of objections to the content and methods of his teaching. (We have Peano's own word that he did not ask to be given that course.)

What Vacca especially remembered from the years of collaboration on the *Formulario* were "the long discussions of the classical texts, from Archimedes, Euclid, Apollonius, up to the most modern, Gauss, Dirichlet, Weierstrass, Dini, Cantor."

In the meantime, Peano had been consulted by the directors of the national mutual cooperative fund for pensions, and although he had never studied actuarial questions before, he was able to master the subject so as to make a report (published 4 June, 1901) that gave the organization a new lease on life [111]. It is one more example of his versatility. Prof. Bagni, minister of agriculture, industry, and commerce, had already made plans to dissolve the fund, as a result of an adverse report from his own actuary. Now he attacked Peano's plan, claiming to have found errors in the mathematics, but he was quickly proved wrong, and the fund was allowed to continue. Moreover, it adopted several changes suggested by Peano.

The report of the commission, of which Peano was a member, was made on 10 August, 1901 [114] and Peano published a reply to Bagni's criticism in a bulletin of the ministry of agriculture, industry, and commerce [115]. The following year (on 12 May, 1902) there was a second report of the commission [120] and on 12 December, 1902, Peano published a note pointing out certain errors in a law, proposed in the senate, to regulate such funds [123]. A pamplet of 18 April, 1903 made some suggestions for the distribution of pensions [124]. The series of publications dealing with actuarial mathematics closed on 15 August, 1905, with a complete project for improving the national mutual fund [131], or rather, it ended with a second edition [111'] of the original publication of 1901 (with a new preface dated 3 November, 1905.)

Returning to the year 1901, we see by the 15 July issue of the *Rivista* that work continued on the *Formulario* project. This issue contains a long list of "additions and corrections" attributed to: E. Cantoni, C. Ciamberlini, G. Eneström, A. Padoa, G. Peano, A. Ramorino, O. Stolz, and G. Vacca [112]. The list concluded with a bibliography of sixty names that refers only to this list. A note on logarithms, countersigned by Peano, may be mentioned as showing his recurrent interest in binary numeration: "If binary numeration is adopted, the corresponding tables of logarithms are five times shorter. The length of the logarithm rules is also reduced in the same ratio." That same issue also contains a very favorable review of an arithmetic text of O. Stolz and I.A. Gmeiner [113], in which Peano makes a comparison with Volume 3 of the *Formulaire*.

From 17 to 22 August, Peano was in Livorno for the second congress of teachers of mathematics in the secondary schools. Peano arrived with a proposal that the congress support the compilation of a dictionary of mathematics. Typically, he was armed with an eight-page provisional version of the part on mathematical logic [116], that he immediately presented to the congress. The proposal was accepted by the congress, gaining also the support of Mathesis, which had organized the congress, and the journals *Periodico di Matematica* (Lazzeri), *Rivista di Matematica* (Peano), *Il Bollettino di Matematica* (Conti), and *Il Pitagora* (Fazzari). (With additions by Vacca, Vailati, and Padoa, this was published in the 4 December issue of the *Rivista* [116'] and printed as a separate sixteen-page pamphlet on 9 December [116"].)

That fall, Gino Fano, who graduated from the University of Turin in 1892, returned to Turin, having been named Extraordinary Professor of Projective and Descriptive Geometry on 5 August, 1901. Rodolfo Bettazzi (one of the founders and the first president of Mathesis), having taught since 1892 at the Military Academy in addition to his secondary school teaching, now added an introductory course at the university on the theory of functions, having just obtained the *libera docenza* in infinitesimal calculus.

The final number of Volume 7 of the *Rivista* appeared, dated 4 December, 1901. It contained further "additions to the *Formulaire*" [117], contributed by: A. Arbicone, T. Boggio, E. Cantoni, F. Castellano, G. Peano, and G. Vacca. One of the additions, countersigned by Peano, accepts one of Lindemann's purported proofs of Fermat's last theorem. Set in boldface type is the remark: "Thus the conqueror of π has also conquered Fermat's last theorem!"

There was one other publication in 1901, a letter to the editor (Alberto Conti) of a journal for elementary school teachers [116.1]. The letter, which is dated 10 July, 1901, treats a small matter, but is of interest for the reaction it produced. The editor follows it with: "In view of the new aspect assumed by the question mentioned by Prof. Peano, in view, above all, of the authority of the name of the illustrious professor of the University of Turin, we have hastened to publish his letter and to consult one of our colleagues in the matter." Thus, while Peano's influence among his colleagues at the university may have been decreasing, the schoolteachers still respected and listened to him. It was perhaps this that prompted him to think that his ideas and methods might be introduced into the schools. To this end he extracted from the *Formulaire* the propositions dealing with arithmetic and elementary algebra, translated the explanatory notes into Italian, and presented the volume for possible use in the secondary schools [118]. The book opens with a comparison of its contents with the official mathematics program (1902) of the upper *ginnasio* and the *liceo*. It is dated only '1902', but must have appeared sometime after February, as there is a note in the preface saying that the fourth edition of the *Formulaire* was already being printed. The first 224 pages of this edition were printed the week of 17–22 February. As before, this edition was printed in Turin and set with the type of the *Rivista* at Peano's villa in Cavoretto. For this edition, Peano supervised the typesetting (all done by hand) and often set type himself. The actual printing was done at the Tipografia Cooperativa in Turin.

In the first number (13 March, 1902) of Volume 8 of the *Rivista* there is a review, written by Mineo Chini, of a calculus text by Cesare Arzelà. Peano followed this with a "confrontation with the *Formulario*" [119]. Of interest is Peano's continued rejection of the Axiom of Choice, but it is ironic that one of the definitions of Arzelà that Peano refers to as a 'possible definition' was later pointed out by Cassina to be 'possible' only if the Axiom of Choice is admitted!

On 20 March, Peano bought two more plots of land in Cavoretto.

In a letter, dated 23 May, 1902, to L. Gérard, editor of the *Bulletin des sciences mathématiques et physiques* (and published in the issue of 15 June [121]), Peano discusses the definition of imaginary numbers. Once more we see his interest in mathematical definitions. He wrote:

A further remark relative to the form of the preceding definitions: Logicians, following Aristotle, divide definitions into real and nominal. All mathematical definitions are nominal. They give the rule that definitions proceed 'per genus proximum et differentiam specificam." No mathematical definition satisfies this rule.

Ernesto Laura graduated that year 'with high honors' in mathematics. Peano chose him to be assistant for calculus (Vacca had moved to Genoa, but continued his collaboration on the *Formulario*), thus reviving D'Ovidio's custom.

In 1898, Peano had presented a paper to the Academy of Sciences of Turin that developed the theory of vectors on the basis of three undefined terms: *point*, the relation of *equidifference* among four points (which leads to the notion of vector), and the *internal* product of two vectors. In a series of notes beginning in 1896, Mario Pieri showed various ways in which geometry can be based on only two undefined terms, and in one of the papers presented at the International Congress of Mathematicians in Paris, he based Euclidean geometry on the notions of *point* and *distance*. Peano combined the two ideas in November 1902 in a brief note for the Academy of Sciences [122], i.e., he showed how to base his theory of vectors on these two primitive terms.

With the printing of pages 225–312 of the fourth edition of the *Formulaire* from 30 April to 3 May, this edition appeared to be essentially complete, but it lacked a preface, index, and bibliography. Before this could be prepared, the need was felt to include "additions" (which, when finally printed, occupied 54 pages, so that the volume was not completed until early 1903 [125]. (The remaining parts were printed on 12–13 March.) This volume differed from the earlier ones in several ways. First, there is a slight change in the name (from *Formulaire de Mathématiques* to *Formulaire Mathématique*.) Its content of mathematical logic is essentially that of the third edition, but now this is not presented in a single section, but introduced as needed for the various mathematical sections. The mathematical content has been increased with the addition of other areas of mathematics as well as additions to areas presented in the earlier editions. It has almost twice as many pages. Peano, when he gave a copy to the Academy of Sciences on 5 April, described it as containing approximately 10,000 formulas.

In his preface, Peano gives the names and addresses of the various collaborators. He singles out Vacca for special mention: "Mr. Vacca has once again greatly aided us with this volume, as with the preceding volumes 2 and 3." The preface is followed by seven pages of "exercises of mathematical logic," to help those "who already know mathematics" to read the *Formulaire*. Vacca prepared the 17-page bibliography and the index of names. This is followed by a 14-page 'mathematical vocabulary,' which complements the 'dictionary of mathematical logic,' presented to the congress of secondary teachers in Livorno in August 1901 [116]. In the preface to it we find the first hint of future developments in the lone sentence: "The words are written in Latin, if this does not differ greatly from the French." Peano does not say here why he does this, but the answer was to come shortly and would open up a new phase of his activity — one that would last for the remainder of his life.

Peano's interests were beginning to shift, but Bertrand Russell, whose interest in the matter had been inspired by Peano, that spring sent Peano a copy of his book *The Principles of Mathematics*. Peano immediately recognized its value and in a 'thank-you' note of 27 May, 1903, he wrote that the book "marks an epoch in the field of philosophy of mathematics."

CHAPTER 15

LATINO SINE FLEXIONE

In the 20 October, 1903, issue of the *Rivista* (Vol. 8, No. 3 — there was no No. 2!) appeared Peano's proposal for an international auxiliary language [126]. (A preliminary version had been printed separately on his birthday, 27 August [127'].) As we have seen, the idea of such a language was already 'in the air' and Peano made no claims to originality for his idea. On the contrary, he gave all credit to Leibniz. His method of presenting it, however, was a stroke of genius. It is a refinement of his habit of not making a proposal until it has already been, at least partially, put into practice.

The language Peano proposed was a very simplified form of Latin, Latin stripped of all grammar — a Latin without flexions, or *Latino sine flexione*. His article describing it begins in classical Latin and, one by one, as suggestions are made for simplifications, they are immediately adopted, so that the article closed in Latino sine flexione. Most of these suggestions are found in the writings of Leibniz, and Peano gives the appropriate quotation at the beginning of each section. After the simplications have been made, the remainder of the article, written entirely in Latino sine flexione, gives a brief history of artificial languages.

One sentence is especially interesting in the light of future developments: "The question of an *inter-lingua* at this time has nothing in common with the ideography that we use in the *Formulario mathematico*." Later, he would propose the name 'Interlingua' for his academy. The fifth, and final, edition of the *Formulario* would be called '*Formulario Mathematico*' and its language would be Latino sine flexione.

This note is followed by another [127] containing several statements in German and English, concerning the 'principle of permanence' (in defining new mathematical entities), and their translation into Latino sine flexione.

It is not clear just when Peano's acceptance of Latino sine flexione as a solution of the question of an international auxiliary language occurred. It seems clear that this took place before the spring of 1903. This is shown by the use of Latin versions of words in the mathematical vocabulary appended to the fourth *Formulaire* (in the part printed on 13 March.) Govanni Vacca claimed that Peano developed this idea after he (Vacca) investigated the unpublished Leibniz manuscripts in Hannover in the summer of 1899. (A brief account was published that year in a journal edited by Gino Loria and some of Leibniz formulas were used in the second *Formulaire*, No. 3 of 1899 [106].) We recall that it was Vacca's meeting with Louis Couturat in Paris in 1900 that prompted the latter to go to Hannover to make a collection of the previously unpublished writings of Leibniz, which was published in a large volume in 1903. Couturat was already preparing a work on the 'logic of Leibniz,' the writing of which was influenced by that visit. This work appeared in 1901 and contains Leibniz' suggestions for an artificial language, quoted by Peano in his article. Thus, it was probably about this time that Peano began to see a simplified form of Latin as a solution to the problem of an international language.

It is possible that Peano became interested in the *problem* much earlier; it was certainly 'in the air'. Padoa was active in the discussions about it at the Paris congresses and Zamenhof, the author of Esperanto, had already written to Charles Méray on 30 May, 1900, that "the publication of a complete dictionary of mathematics will be very useful, and so I wish that the new work may turn out so perfect that it will be selected for the translation mentioned, all the more seeing that Prof. Peano is a very able Esperantist." Zamenhof was, perhaps, writing from hearsay. He probably did not meet Peano until 1906.

The most successful artificial language up to this time was Volapük, but it was already on the decline. Esperanto now seemed the most promising. Peano was elected president of the Volapük Academy in 1908, but immediately transformed it into a 'free discussion' academy. It is doubtful whether he ever seriously considered Volapük a good choice for an international language, but there was a Volapük Club in Turin as early as 1888 and he may have been familiar with it for some time. Some have suggested that Peano's interest in an international language was shown in 1889, that the publication of the *Arithmetices principia* [16] was an attempt to use Latin as an international language. It seems clear, however, that Peano had other reasons for using Latin, though what these reasons were he never said, for never did he refer to the *Arithmetices* as an attempt to use an international language.

The artificial language movement was strong from the turn of the century up to the First World War. Many of its activities were revived after the war, only to be killed off by the Second World War. There was never a real revival afterwards. Latino sine flexione shared this fate, although some use of it may still be found. Esperanto, because of its strong humanitarian identification, has always had a loyal corps of adherents, but it is unlikely to ever really be successful.

Peano's only other publication in 1903 was a note in the *Intermédiaire des Mathématiciens* [127.1] asking for the origin of the letter C turned upside down used as a symbol for deduction. (He mentions its use by Gergonne and Abel, which had been pointed out by Vailati.)

That fall, Tommaso Boggio added being assistant for calculus to his other duties. Vacca was still in Genoa and was socialist municipal councilor and member of the national administration of the socialist party, as well as assistant in mineralogy at the University of Turin, and according to his own report later, he attended many of Peano's lectures. He also gave a course in mathematical logic at the University of Genoa! Vacca was one of the first converts to Peano's Latino sine flexione; the very isssue of the *Rivista* in which Peano's article appeared also contained a note by Vacca written in this language.

The University of Turin dates its beginning from 27 October, 1404, the date of a papal bull of the antipope Benedict XIII (the Spanish cardinal, Pedro de Luna), decreeing the erection of a university in Turin. Thus, in the fall of 1903 the university began the celebration of its 500th anniversary. Appropriately, the academic year opened with student disturbances.

At Innsbruck University the previous 16 May fights broke out between the German-Austrian and the Italian-Austrian students. This revived a demand for an Italian university in Trieste, which the Austrian government, however, refused to consider. Then on 23 November the Italian Law Faculty opened in Innsbruck, but more serious riots occurred, involving the townspeople, and the authorities closed it the same day. Naturally there were demonstrations in the Italian universities. The rector of the University of Turin wrote in his report for the year:

The tranquility of our studies was, although slightly and for a short time, disturbed once throughout the faculties, and another time limited to the course in general chemistry. The motive for the first was the sad happenings at the University of Innsbruck, so deeply offensive to our nationality; but the demonstrations quickly ended, classes were immediately resumed, and, with no more interruptions, proceeded normally for the remainder of the year. As for the second incident, which caused the suspension of the course in general chemistry, the cause was the deplorable act of lack of discipline on the part of the students; but steps taken by the administration kept this also to a minimum.

More unfortunately, the festive spirit of the anniversary year was dampened by a fire which destroyed the National Library that was also associated with the university. Some 24,000 books and other important objects were lost, including 2,600 manuscripts.

On 3 January, 1904, Peano read his paper on 'Latin as an international auxiliary language' [128] at the Academy of Sciences. This paper is in Italian, but has examples of Latino sine flexione, and Peano asked to have his *Rivista* article reprinted in the *Acts* of the Academy. (This was not done, however.) By now Peano was very much *au courant* with the international auxiliary language movement and he described several of the latest proposals for an international language.

On 17 January, having excused his absence, he asked D'Ovidio to present to the Academy for him an article of Mario Pieri. Peano continued to support Pieri in this and other ways. For example, Pieri had entered the competition for the third Lobachevskii Prize to be given by the Physico-Mathematical Society of Kasan and it was Peano who wrote the report on his work [129]. (This competition was founded by international subscription, on the initiative of A.V. Vasil'ev. The first prize was awarded in 1897 to Sophus Lie.) In this case, the decision was made (on 27 February, 1904) to award the prize to Hilbert for the second edition (1903) of *Die Grundlagen der Geometrie*. Pieri received 'honorable mention', along with Paul Barbarin, Emile Lemoine, and Eduard Study. (Poincaré received a gold medal for writing the report on Hilbert, while the other referees, including Peano, were made honorary members of the Physico-Mathematical Society.) Peano closes his report with a quotation from Bertrand Russell's *The Principles of Mathematics*: "This is, in my opinion, the best work on the present subject." Peano seems to imply that this referred to Pieri's work in general. In fact, Russell was referring in particular to the memoir on 'The principles of the geometry of position,' published by the Academy of Sciences of Turin in 1898.

In his paper on 'Latin as an international auxiliary language' [128], Peano had distinguished Latino sine flexione from Latino *minimo*, the first essentially achieved, the second yet to be constructed. He had shown, in theory and by example, that the best grammar was no grammar. He now proceeded to the second task of discovering the minimum vocabulary of this emerging language. From the beginning his goal was a vocabulary that was already international (at least in western Europe), i.e. the words were not to be manufactured, but discovered – and the way to do this was to make a comparative study of the European languages to find out which Latin words were current. The first result of this study was a forty-page pamplet [130] 'A vocabulary of international Latin, compared with English, French, German, Spanish, Italian, Russian, Greek, and Sanscrit.' In 1909 this would be expanded to a booklet of 87 pages [141] and in 1915 to a volume of 352 pages [169].

In the meantime, he determined that the fifth edition of the *Formulario* would be in Latino sine flexione. Thus, besides greatly increasing the mathematical content of this volume, he had the task of translating all notes into Latino sine flexione, and this required a prior determination of the form of the words that would be most internationally understood. A forty-page 'Index and Vocabulary' to this edition appeared in 1906 [137], but, although the first 272 pages (about half the book) were printed by 10 November, 1905, the complete edition did not appear until June 1908 [138]. Regarding Peano's marginal notes in his own copy, Cassina has remarked that they were intended to serve for an eventual future edition. It seems unlikely, however, that Peano ever seriously considered a sixth edition, and that this was simply a continuation of his habit of making marginal notes. (As Cassina has also noted, Peano sometimes had his books bound with interleaved blank pages.) The last issue of the *Rivista* appeared in 1906 and, as we shall see, Peano's interest shifted dramatically in 1908.

The Second International Congress of Philosophy was held in Geneva, 4–8 September 1904. Peano and Vailati, both members of the permanent international commission elected at the first congress in 1900, attended, as did their friend Vacca. (Among the variety of people at this congress was the Russian composer-pianist Aleksandr Skryabin.) Couturat was also there and read no less than four papers. In addition to the sections of the first congress, a section on History of Science was added at the request of Paul Tannery. (He died shortly after the congress ended and a biography by his brother Jules was included in the published report of the congress.) This section met jointly with the Congress of History of Science, which counted this their third congress, the second having taken place in Rome, 1–9 April, 1903. (Vacca and Vailati had attended the Rome congress, but not Peano.)

Peano did not present a paper at this congress, but did take part in several discussions with his typical ability to uncomplicate situations. For example, A. Naville, of Geneva, in a talk on the notion of law, had given as an example the law of gravitation, and implied that Kepler's laws were not really laws. In the discussion that followed, Peano commented to the effect that the law of gravitation and Kepler's laws imply one another, that one can be logically deduced from the other, and that therefore they have the same character of necessity. Again, Pierre Boutroux talked on the notion of correspondence in mathematics. Peano rose to clarify the distinction between logical relation and mathematical function. When Boutroux tried to twist his words, Peano responded that the theories were distinct, but one implied the other. Russell, he said, had defined mathematical functions by means of logical relations, while he, in the *Formulaire*, had taken the opposite direction. "There is neither separation nor opposition between logic and mathematics, but a logical passage from one to the other."

Louis Couturat, in his report on the progress of the idea of an international language, recalled that the delegates of various congresses (he representing the First International Congress of Philosophy) met on 17 January, 1901, to discuss the question. At that meeting a program was adopted and the *Délégation pour l'adoption d'une langue auxiliaire internationale* was founded, with Leau as secretary and Couturat as treasurer. He also reported that 200 organizations supported the *Délégation* and proposed that the Congress vote its support of the *Délégation*, renew his mandate as delegate from the Congress, and select a new (preferably non-French) delegate. His proposal was adopted unanimously and Ludwig Stein, of Berne, was selected as the new delegate. Peano, Vailati, Calderoni, and Cantoni were all re-elected to the permanent international commission and the Congress closed with a banquet in the Grand Theatre, offered by the Canton and City of Geneva.

In the fall of 1904, Vacca returned to Turin as Peano's assistant. At the age of 46, Peano was now 33rd in seniority in a total of 48 'ordinary' professors at the University. Peano continued to assist his friends by furnishing an outlet for publication in the *Acts* of the Academy of Sciences. For example, on 8 January, 1905, he presented, for insertion in the *Memoirs* of the Academy, a paper of Mario Pieri. As papers intended for the *Memoirs* were not automatically accepted, the president delegated Peano and Corrado Segre to report on it. Their report [131.1] was read by Segre on 5 February. It was favorable, and insertion in the *Memoirs* was unanimously approved. This paper gives a postulate system for complex projective geometry, without deriving it from real projective geometry. The 'reporters' remark that "the Memoir of Prof. Pieri will have the merit of having been the first to resolve this interesting problem." During this academic year, Peano presented five papers of other authors, and none of his own.

Although Peano enjoyed generally good health, in a note to Bertrand Russell on 7 March, 1905, he complained: "My health is not very good." There was no explanation of what the difficulty might be.

On 12 July, 1905, Peano was elected a Corresponding Member of the *Accademia dei Lincei*. (President Blaserna communicated the thanks of the newly elected, Peano among them, at the session of 5 November.) This is one of the highest honors for an Italian scientist, this venerable academy (founded in 1603 by Prince Federico Cesi) having enrolled Galileo as a member during his visit to Rome in 1611.

Vacca having left for Florence to study Chinese, Peano named Giuliano

Pagliero as assistant for calculus. Pagliero held this position for fifteen years, the longest period of all of Peano's assistants. Needless to say, he shared many of Peano's views. He collaborated on the *Formulario* project and in 1907 he published a book on applications of calculus written in Latino sine flexione. Pagliero was Peano's assistant until 1920 and their contact continued after that, although Peano reported to one of his correspondents in 1929 that Pagliero was "totally occupied with his numerous family."

Peano's first note for the Academy of the *Lincei* since becoming a member was received at the session of 7 January, 1906, [132]. It is a very brief statement of four theorems concerning finite differences, that were probably discovered in his actuarial work discussed earlier. They were published here for the first time, but had, in fact, already been printed as part of the *Formulario Mathematico*, Vol. 5 (which was not completed until 1908.)

CHAPTER 16

THE COTTON WORKERS' STRIKE

Peano was never active politically, but his socialist sympathies were well known. This is illustrated by an incident that occurred in the late spring of 1906. On Wednesday, 2 May, of that year about 900 cotton mill workers (almost all women) at the Cotonificio Bass in Turin requested a reduction of the work day from 11 to 10 hours. The company refused this request and the workers went on strike. When the workers at the Cotonificio Hoffmann (about 600) heard about this the next day, they immediately left their work in an act of solidarity that became the signal for an almost general movement. In fact, part in the morning and part in the afternoon, workers in the other important factories left their work: Poma (about 1500), Mazzenis (about 900), Wild and Abegg (about 700), so that by the end of the week about 12,000 women were on strike.

On Saturday there were several demonstrations and the workers formulated the following demands for presentation to the owners:

(1) Normal work day of 10 hours.

(2) Interval for lunch of 2 hours in summer and 1½ in winter.

(3) Five minutes grace period on entering and leaving.

(4) That the fines, not for breakage, but for simple infractions of the rules be put in an account to be administered by workers in accord with management, for the benefit of the sick and of accident victims.

(5) A relative increment so that the current wages would not be reduced, and this for men as well as women, for night as well as day work.

(6) For the garment factory workers, abolition of the so-called *spese* (costs for needle, thread, etc.) as well as the deduction for breakage not caused by carelessness or for use of the machine.

But, despite the urging of Senator Frola and Prefect Gasperini, the owners refused to bargain. The strike leaders then called for meetings of the strikers at 9 a.m. and again at 2 p.m. on Sunday. In the meantime, according to the newspaper *La Stampa*, "a gentleman who owns a villa in Cavoretto, in sign of sympathy for the workers, has sent to them, through the *Camera del lavoro*, an invitation to an outing in the nearby region. The workers joyously accepted the invitation and will go there in procession following the 2 p.m. meeting."

On Sunday the strike situation had not changed; the most notable event was the outing to Cavoretto. (What follows is the account of this in *La Stampa* of 7 May 1906. The reporter is obviously not sympathetic, but does have the merit of being an eyewitness.)

The Outing to Peano's Villa

Yesterday morning we gave the news that a gentleman, in sign of sympathy for the strikers, had invited them, though the *Camera del lavoro*, to make an outing to his villa in the region of Cavoretto. Later we learned that the gentleman making the invitation was Prof. Peano, professor of infinitesimal calculus in our university, who is the owner of a small villa in Val Pattonera, about an hour's walk from the entrance to Cavoretto.

The invitation was greeted by the workers with noisy demonstrations of joy. The prospect of a lovely hike in the hills with numerous companions presented itself to their minds as a very attractive novelty.

Certainly the episode constitutes a festive intermezzo that is entirely original and without precedent in the annals of Turin strikes.

This explains why there were so many girls gathered at 2 p.m. in the General Insurance building in order to take part in the curious outing. Of course, it was preceded by several speeches, for it is well known — and the Socialists know it better than anybody — that speechmaking constitutes the driving force of all collective manifestations.

Toward 3 p.m. the hikers went out into the main street Corso Siccardi and, preceded by the majority of the *Camera del lavoro*, wearing red ribbons, they formed a procession. Naturally, the women were joined by an equally large number of men, making altogether a group of about 4000 people.

Among the hikers were city councilors Del Bondio, Sciorati, Daghetto, Castellano, Alasia, and provincial councilors Barberis and Viglongo.

The procession went through Corso Siccardi, Corso Vittorio Emanuele, Corso Massimo D'Azeglio and, by way of Corso Dante, arrived at the street to Cavoretto. Along this path the strikers sang the 'Hymn of the Workers,' giving way between the verses to loud laughter, in which there was the exuberance of youth, the noisy gaity of that happy age.

The passage of this group in procession constituted a curious note for the numerous citizens who were enjoying themselves along the holiday afternoon streets.

As they went along, a swarm of young girls were distributing numerous small red sheets with the manifesto of the *Camera del lavoro*.

Up to the steps leading to Cavoretto and even for a piece of the road leading to Val Pattonera, the joy of the hikers — expressed in merry shouts and lively popular songs — grew with a Rossini crescendo; but when the climb began to be more steep and the sun's rays more painful because of the fatigue of the unusual exercise, their voices began to be muted, they laughed less, and everyone concentrated his thoughts on the desire for a liquid and possibly cool refreshment.

Toward 5 p.m., after two hours of march, the first ranks of the procession finally reached Peano's little villa, almost hidden among the other cottages and by a thick growth of trees. The professor was there, waiting for his guests and happily smiling to see that they were so numerous.

But the hikers no longer appeared equally happy: fatigue and thirst dominated all. The exclamations, the questions, the answers loudly interrupted one another.

"Are we finally there? Where is this villa? Don't you see it? It's here. But where can we go? There, down there in that field. Isn't there anything to drink? Yes, yes, down there, look around, there's some water. I'm really thirsty. Move on! Move on!"

And the hikers hurried down in a tumult, cutting across fields, jumping ditches, leaping hedges, all dominated by an irresistible desire to find a bit of water to quench their thirst.

The scene was certainly curious: certainly, never had the inhabitants of that lonely spot seen a more numerous and noisy wave of people climbing over their rustic garden plots.

Finally some volunteers succeeded in finding a pail and a well, and among the cries of hosanna from the crowd the work of thirst-quenching began. In that moment as never before we understood the sublimity of the precept contained in the work of mercy expressed by the Divine Master in the words: Give drink to those who thirst!

Meanwhile, a good number of hikers, thirstier than the others, had left the green hill and were hurrying at full speed for Cavoretto in search of beer or wine, while others had thrown themselves down, tired, under the friendly shade of the trees to refresh themselves a bit with the cool breeze. After a half hour of rest, the remaining gathered in a field to hear more speeches. Those speaking were councilors Viglongo, Alasia, Castellano, Barberis.

This last, who had selected a cherry tree as podium, spoke in Piedmontese dialect, and a few phrases of his speech have remained in my mind:

"Instead of getting drunk like animals, it seems to me that the workers should often breathe the oxygen-rich air of these hills. You women especially ought to tell your men, and when they come home drunk you ought to take your broom and give them a few. . . ."

Barberis found himself called down by Commissioner Gualti who, with a good number of guards and *carabinieri*, had followed the hikers.

The last to speak was the anarchist from Romagna, Massetta, who of course defended the general strike and urged an energetic 'popular action', read: breaking of windows and worse if . . . possible!

But the inflammatory words of Massetta were opposed by Viglongo, who said that the working class had in solidarity a weapon to fight for its cause more effective than disorder in the streets.

Toward 6:30 p.m. the last hikers left the hill to descend precipitously the serpentine streets.

Alas! Just at that moment arrived on the spot a cart loaded with wine and food that some volunteers had gone to buy in Cavoretto!

But by now the hour was pressing and the first signs of a coming shower were appearing in the sky, giving the legs the energy and the mind the desire to quickly reach the main street to catch the tram. And every striker was happy at that moment that among their striking colleagues the tram drivers were not included!

The next day (Monday 7 May) the number of strikers had reached more than 20,000. The strike was finally settled on the following day with a reduction to 10 hours at the same pay. Despite the damage done to his garden, it is said that Peano never regretted this gesture.

CHAPTER 17

COMPLETION OF THE *FORMULARIO*

On 31 March, 1906, Peano sent an article to the *Circolo Matematico* of Palermo that was reported at the 8 April session and printed in the *Rendiconti* [133]. It shows that he had not lost interest in questions pertaining to the foundations of mathematics and, on the contrary, had some penetrating things to say. It is true that he had been somewhat reticent. He never commented in print on Burali-Forti's discovery of an antinomy in Cantor's set theory in 1897 and Bertrand Russell, after discovering his paradox a year earlier, complained is his 16 June, 1902, letter to Frege: "I have written to Peano about this, but he still owes me a letter." Peano did, however, comment on some of the ideas of this article in a letter to Russell of 16 February, 1906.

Peano's article (written in Latino sine flexione) is on the Cantor-Bernstein theorem: If x and y are cardinal numbers such that $x \leqslant y$ and $y \leqslant x$, then $x = y$. He points out that G. Cantor had published this in both the *Mathematische Annalen* and the *Rivista di Matematica* in 1895, without proof, and that a proof by Felix Bernstein was published in Borel's *Théorie des fonctions* in 1898. Peano then proceeds to give an entirely different proof. This proof is in a letter that Dedekind wrote to Cantor on 29 August, 1899, but did not publish. It was also independently discovered by Zermelo and published by him in 1908.

In this article, Peano also takes the occasion to produce a model in set theory of his postulates for the natural numbers. He then remarks: "Thus it is proved (if proof were necessary) that the postulates of arithmetic, which the collaborators of the *Formulario* have shown to be necessary and sufficient, are not self-contradictory." Then, after noting that other models had been given by Burali-Forti and Russell, he gives us a statement of his views on the foundations of mathematics that clearly shows that he was not a 'formalist':

But a proof that a system of postulates for arithmetic, or for geometry, does not involve a contradiction, is not, according to me, necessary. For we do not create postulates at will, but we assume as postulates the simplest propositions that are written, explicitly or implicitly, in every text of arithmetic or geometry. Our analysis of the principles of these sciences is a reduction of the ordinary affirmations to a minimum number, that which is necessary and sufficient. The system of postulates for arithmetic and geometry is satisfied by the ideas of number and point that every writer of arithmetic and geometry has. We think number, therefore number exists.

A proof of consistency of a system of postulates can be useful, if the postulates are hypothetical and do not correspond to real facts.

On 24 July, 1906, Peano wrote Bertrand Russell to thank him for a copy of Russell's article 'The theory of implication' that was published that year in the *American Journal of Mathematics*. Apparently in answer to a question of Russell, Peano comments on the article just mentioned. He indicates that he plans no further publication on this matter and tells Russell to use any of his ideas.

Nevertheless, this article was reprinted (with only slight change) in the last issue of the *Rivista* (24 August, 1906) [133'], with a supplement. He begins the supplement with a bibliography of recent articles dealing with antinomies in set theory and says that "every antinomy, ancient or modern, depends on a consideration of the 'infinite'." "The advice given by some authors," he continues, "not to consider the 'infinite', is prudent, but does not resolve the problem, for infinity is in the nature of many questions, and '*naturam expelles furca, tamen usque recurret*'." ("You may chase nature out with a pitchfork, it always keeps rushing back.") The quotation is from the Roman poet Horace, and is a rare poetic moment for Peano!

Peano once again objects to the Principle of Zermelo, remarking that if the postulate of Zermelo is used in the proof of a theorem, then that proof is not valid according to the ordinary value of the word 'proof'. He then shows how to avoid this postulate in proving that the real numbers are uncountable, using Cantor's diagonal method. This leads into a discussion of a paradox that Jules Richard published in the 30 June, 1905, issue of the *Revue générale des sciences pures et appliquées* (which was also discussed by Poincaré in 1906.) This paradox may be stated as follows: Let E be the class of all non-terminating decimals that can be defined in a finite number of words. Since the definitions can be arranged lexicographically in a sequence, E can also be formed into a sequence. Now let the nth digit of the nth member of this sequence be altered systematically according to some definite rule — say by being increased by 1 unless it happens to be 8 or 9, and by being replaced by 1 in these exceptional cases; and let the digit so obtained be taken as the nth digit in a new decimal N ($n = 1, 2, \ldots$). Then N differs in at least one decimal place from every member of E, and it therefore does not belong to E; but it is nevertheless a non-terminating decimal, defined in a finite number of words.

Peano rejects both Richard's own solution of the paradox and "the true solution" of Poincaré. Indeed, he remarks that "the objection to the solution of Poincaré is so obvious, that I doubt whether I have understood the author well." He then gives a "new solution" to the paradox, and concludes with the very penetrating remark:

The example of Richard does not pertain to mathematics, but to linguistics. One element, which is fundamental in the definition of N, cannot be defined in an exact fashion (according to the rules of mathematics.) From an element that is not well defined may be deduced many self-contradictory conclusions.

This article (dated 23 August, 1906) was printed the next day. Peano posted a copy to Russell and left almost immediately for Geneva to attend an Esperanto congress. (At this congress Peano met L.L. Zamenhof, creator of Esperanto, for the first time. When they were introduced by A. Michaux, Zamenhof said with a smile, "If my disciples see me now, they will excommunicate me." Peano replied, "I have few disciples, but they are all tolerant: it is one compensation.") But just before leaving, he received from Couturat the page proofs of Russell's article on this subject. Peano returned to Turin on 5 September. Russell wrote him the next day and Peano received the letter just in time to write (on 9 September) before leaving the following day for the South, where he planned to spend the remainder of September (visiting Modena and the military colleges in Naples and Rome.) He praises Russell's article, saying that if he had received it sooner, he would have had no need to publish his own article in the *Rivista*.

This last issue of the *Rivista* concludes appropriately (symbolic of Peano's shift of interest from mathematics) with 'News of the international language' [134]. His optimism is obvious. He wrote: "The question is of general interest, and I give several examples of new projects for an international language, the first five being taken from *La Revue de l'Esperanto* of A. Michaux. They have different names, though the convergence toward a solution is obvious."

The Royal Commission for the reform of the middle schools had asked teachers for their opinions and the *Bollettino di Matematica* had offered to print them. Peano took the occasion to offer some incisive comments on whether or not students should be taught to treat proportions according to the methods of Book V of Euclid's *Elements* [135]. His answer is essentially this, that for practical purposes it can be omitted, since the algebraic treatment must be taught anyway.

In a letter to the editor of *L'Enseignement Mathématique* [136] he quotes a theorem concerning series from the *Formulario Mathematico*, although this edition has not been completely published. This (and a reply to a question in the *Intermédiaire des Mathématiciens* [137.1] quoting another theorem) show that there had already been some distribution of the parts (about half) that were printed in 1905. The 'Index and Vocabulary' to the *Formulario Mathematico* was published separately [137].

January 1907 was exceptionally cold and many trains were blocked by snow. In spite of this, Peano visited Zurich this month, probably during the early part. (The university's Christmas vacation lasted through 9 January.)

On 10 February, Peano wrote what became his only publication of the year, and he did not intend it to be published. It was a letter to S. Catania [137.2] commenting on an article Catania had earlier published in *Il Pitagora*, a journal for schoolteachers of mathematics. Catania, however, passed it along to the editor, who immediately published it.

In September, Peano was in Parma for the First Congress of the Italian Society for the Progress of Science. The next month he was in Paris for the meeting of the *Délégation pour l'adoption d'une langue auxiliaire internationale*, which began on Tuesday, 15 October, at the Collège de France. The *Délégation* met twice a day until 24 October, but Peano remained only until Saturday, 19 October, when he had to return to Turin because of the university examinations. Consequently, he did not participate in the voting.

The year 1908 saw the termination of the *Formulario* project and a deeper commitment by Peano to the international language movement. The fifth edition of the *Formulaire*, now *Formulario Mathematico* [138], is dated (in the preface) "June 1908", but half of it was printed in 1905, and the other half earlier in 1908. This edition contains some 4200 formulas and theorems, completely stated (i.e. hypotheses given) and in many cases proved. There is a wealth of historical and bibliographical information and often direct quotations from original authors of theorems are given. As Cassina repeated many times, it is an "inexhaustible mine of science." One *has* to be impressed by the amount of information contained in one volume of such modest size (516 pages) and by the variety of subjects covered: mathematical logic, arithmetic, (classical) algebra, theory of equations, elementary geometry, geometrical calculus, vectors, quaternions, calculus of finite differences, differential and integral calculus, and elementary parts of probability, calculus of variations, and differential equations.

Despite all this, or perhaps because of it, this book received little real use. Although this work is a classic of mathematical literature, in that it is one of the most successful attempts at a compendium of mathematical knowledge, like other attempts, and in spite of some novelty of mathematical results, it was out of date by the time of its publication. Mathematics is a living and growing subject and efforts at encompassing it usually end as monuments to the subject and to their authors.

Peano, of course, used it as a text for his classes, but the ideography in which its formulas are written had long since come to be referred to by his students as "the symbols" — and now these symbols were fossilized. There was another, more paradoxical, reason why the book was not read. Peano's decision to use Latino sine flexione for the commentary and explanatory material was prompted by his desire to show that this was indeed an international language. In fact, most mathematicians were put off by the strange appearance of the language and made no attempt to read it. It almost certainly would have had a wider reading audience if it had been printed in French, as were the preceding editions. Thus, access to this "inexhaustible mine", already difficult because of its very conciseness (and the symbolic writing which made this possible), was made even more difficult. There is a great deal of irony in this, and not a little tragedy. One wonders how much of this Peano himself realized. In his later mathematical writings Peano often referred to this publication. Few other mathematicians did so.

Before the *Formulario Mathematico* was completely printed, Peano was in Rome for the Fourth International Congress of Mathematicians, 6–11 April, 1908, of which he was a member of the international committee. He did not present a paper of his own, but on Wednesday, 8 April, he read a paper of Pierre Duhem, who was unable to attend, and he took part in several discussions, notably the discussion following Emile Borel's talk on the teaching of mathematics in the secondary schools of France, and that following Roberto Marcolongo's talk on the possibility of unifying vectorial notation. Peano's rejection of Zermelo's Postulate did not prevent him from having a friendly conversation with Ernst Zermelo. The latter wrote to his friend Georg Cantor on 24 July, 1908: "You missed little by not going to the congress in Rome, since the usual magic of a visit to the Eternal City was spoiled by the terrible weather. Besides, the arrangements were such that it was scarcely possible to meet interesting personalities. Only with Peano and Russell, both of whom are very affable, was I able to talk at some length."

There were six graduates in mathematics this year from the university, of whom five were women, and they were all given diplomas from the Teachers' School as well. It had been only a few years since the first woman graduate in mathematics, but their number increased and continued to be an important part of the mathematics students at the university. After Pagliero's long term, Peano had only women assistants (i.e. from 1920.)

At the end of the summer of 1908, Peano went to Heidelberg for the Third

International Congress of Philosophy (1–5 September). He was a member of the permanent international commission, as was his friend Vailati, who also attended. Vailati may have been part cause of so much of the congress being given to a discussion of pragmatism. He was not, however, re-elected to the permanent commission. Nor was Peano.

The principal address at the first general session was given by Josiah Royce of Harvard University. Also attending from the United States were Paul Carus, editor of *The Monist*, and Fabian and Christine Ladd Franklin. Louis Couturat was there, and his views were of course supported by Peano. He talked on an application of logic to the problem of an international language. He noted in his talk that the *Délégation*, to which he was the representative of the congress, had fixed the principles of the "definitive auxiliary language" and had practically realized it. He did not *name* the language, however, referring to it only as "the international language." Prof. Staudinger (from Darmstadt) rose to point out that the *Délégation* had selected a simplified Esperanto. Peano publicly congratulated Couturat on the talk, remarking that there were many connections between logic and the problem of an international language. Significantly (a hint of future developments), he added that it would be a good idea to form an organization, or academy, that would concern itself with these questions. Couturat replied that the central committee of the *Délégation* would no doubt form the kernel of the academy desired by Peano, and that this would doubtless be done shortly. Peano kept silent, but this was not the academy he had in mind and, as we shall see in the next chapter, he was already making plans to form his own academy.

In October 1908, Peano was in Florence for the first congress of the '*Mathesis*' *Società Italiana di Matematici* (16–20 October) and the second congress of the *Società Italiana per il Progresso delle Scienze* (18–23 October). At the morning session of 19 October of Mathesis, someone suggested dividing the university program in mathematics into two, one for prospective schoolteachers and the other for those seeking a scientific career. Padoa and Beppo Levi were opposed to this, and Peano supported them, noting that even the separation of the engineering program had been regretted, since the majority ended as schoolteachers anyway. The whole question of teaching has been slighted by university professors and ought to be called to their attention, he added. Francesco Severi thought the best way to develop teaching ability was to have a period of practice teaching. Peano pointed out that this had been the custom earlier.

The next day Peano presided at the mathematical section of the congress of the Italian Society for the Progress of Science, at which two papers were read. He then took the opportunity to talk about the existence of an international vocabulary, i.e. the words common to Italian, French, Spanish, Portuguese, English, German, and Russian. He read a long list of such words in the mathematical sciences, pointing out that these were in great part Latin or Greek, that there were only a few Arabic words, and that the influence of the modern languages on the international vocabulary was almost nil. He concluded by noting that no matter which modern language one studied, from Italian to English and Russian, one always met with Latin and Greek.

Despite the objections to his method of teaching, Peano was offered, and accepted, the course in higher analysis at the University of Turin. This course had been taught for twenty years by Enrico D'Ovidio and it was probably D'Ovidio's suggestion that it be given to Peano. He held it for only two years, however, after which time it was entrusted to Guido Fubini.

At the Academy of Sciences, Peano presented, on 15 November, a memoir of Tommaso Boggio. He and Somigliana were delegated to report on it. Peano read their favorable report on 27 December [140] and the section voted to publish it. In the meantime, Peano's plans for promoting an international auxiliary language were maturing, and we must now return to that story.

CHAPTER 18

ACADEMIA PRO INTERLINGUA

When Peano proposed Latino sine flexione in 1903 as an international language he was most concerned with the idea of having a minimum grammar. In this, he took his cue from Leibniz. It was no accident, however, that Latin was chosen for the vocabulary, although Leibniz' use of Latin may have suggested this, too. After that he came to feel more and more strongly that an international scientific vocabulary already existed, but was disguised by the forms peculiar to the various modern languages. That is, he believed that the vocabulary for an international language did not need to be created, but could be discovered in the current usage. He did not deny the possibility of using, say, Volapük or Esperanto, but objected that it was not necessary to invent a new vocabulary. He seemed to feel, also, that people could be persuaded to accept the idea that an international vocabulary already existed, more easily than they could be taught an artificial vocabulary.

This was clearly not a job for one man. In addition to the practical difficulties involved (making word counts, etc.) the language would be more acceptable if it were a result of a cooperative effort. The fate of Schleyer's Volapük was there as a warning of what happens to 'one-man' languages. Ironically, it was from the ashes of Volapük that Peano's academy, phoenix fashion, arose.

Volapük was the first artificial language in modern times to achieve real success. It was published in 1879 by Johann Martin Schleyer, a parish priest in Litzelstetten on Lake Constance, and within a decade it had over a million adherents. The language gained followers in southern Germany at first, but it spread rapidly through France, due to the efforts of Auguste Kerckhoffs, a language teacher in Paris. The first Volapük congress was held on 25–28 August, 1884, in Friedrichshafen, on Lake Constance (Bodensee). Those present were mostly German, but division of opinion was already felt.

Schleyer wanted to be able to translate everything into Volapük, including all nuances of meaning. Kerckhoffs, on the other hand, thought commercial use of the language most important and wanted to see it simplified. He had no real quarrel with its vocabulary, which was primarily English, although deformed in quite arbitrary ways. (The very name Volapük, for example, is derived from 'world-speech'.) Kerckhoffs saw its very elaborate, although regular, grammar as an obstacle. The second congress, with more than 200

attending, was held in Munich, 6–9 August 1887, under the presidency of
Kirchhoff, Professor of Geology at the University of Halle. To answer the
questions that had been raised, this congress founded an academy, the Kadem
Volapüka, having 26 members, with Schleyer as president and Kerckhoffs as
director. Kerckhoffs persuaded the academy to give Schleyer a triple-vote,
but not the veto that Schleyer demanded.

The success of Volapük was phenomenal. By the time of the third congress,
which was held in Paris, 19–21 August 1889, the language had over a million
adherents. There were 283 Volapük societies throughout the world and 316
manuals of the language had been published, 182 of them in the year 1888
alone. That same year there were 25 Volapük periodicals (of which four were
in Italy.) Kerckhoffs was elected president of the third congress, at which, it
is said, even the waiters spoke Volapük. Tension between the president of the
academy and its director continued to grow, however. Schleyer did not
attend this congress, at which Kerckhoffs' faction gained a majority in the
vote for the statutes of the academy.

Still Schleyer demanded the right of veto over the decisions of the academy. The members of the academy then agreed to submit all decisions to the
Inventor, who could then be overruled by a two-thirds majority. Schleyer
insisted on having an absolute veto. In No. 105 of the *Volapükabled* (the
periodical published by him) he suspended the statutes and in No. 114 of
May 1890 he proclaimed new regulations for the academy. This marked a
permanent division in the academy.

Schleyer had moved the previous year into the nearby town of Konstanz
where, in his home on Schottenstrasse 31, he established the *Weltsprache-
Zentralbüro* and published that same year a catechetical grammar of Volapük.
("Who invented Volapük? Schleyer. Where? Lizelstetten." Etc.) Besides his
propaganda efforts for Volapük, he also wrote many hymns, as well as the
epitaph on his tombstone: "I gave mankind a language;/ And pious songs for
the Christian heart;/ My body to earth, my soul to God:/ In heaven we shall
never part." ("Der Menschheit gab ich eine Sprache;/ Den Christenherzen
fromme Lieder;/ Den Leib der Erde, Gott die Seele:/ Im Himmel sehen wir
uns wieder.")

The majority of the academy followed Kerckhoffs, however, and began by
making important modifications in the language. This led to another schism,
as a result of which Kerckhoffs resigned as director on 20 July, 1891. He was
succeeded by a provisional committee. The academy elected its second
director on 14 December, 1892. He was Woldemar K. Rosenberger of St.
Petersburg (now Leningrad).

Although Volapük continued to be used, the academy determined to construct a new language, which was then named 'Idiom Neutral'. Under the third director, Rev. M. A. F. Holmes (of Macedon, New York), elected 14 January, 1898, the academy officially adopted Neutral on 15 May, 1898. From this time the academy was called 'Akademi internasional de lingu universal.'

The academy continued to enlarge the vocabulary of Neutral and made various spelling changes. A number of other changes were suggested, but on 23 January, 1904, the academy established a five-year period of stability for the language, and ceased its activities. This did not, however, prevent individuals from proposing a number of new languages. It must certainly have seemed that the future of the academy was in doubt, that it would be nearly impossible to revive it after the five-year quiescent period. It was the genius of Peano that he saw a source of unity where others saw only diversity. He therefore presented himself in 1908 as a candidate for the academy and, at the same time, a candidate for the directorship. It was a bold stroke, and it worked.

The first Volapük Club in Italy was founded at Piadena (Cremona) in 1886 by Mario Zama. By 1888 there were clubs in many other cities, including Turin, and Augusto Actis, one of the original members of the academy, published a Volapük-Italian dictionary. When Peano published his Latino sine flexione in 1903, he was no doubt aware of Volapük activities, but there is no evidence that he played an active role in any such organization until 1908. By 1903 he was probably aware of, although he did not mention it until later, the work of Julius Lott, of Vienna, who published a language based on a comparative study of the vocabularies of English, French, German, and Italian, and that of Liptay, in Chile, who published a similar language in 1890. At any rate, Peano shared their conviction that an international vocabulary already existed. It needed to be discovered, not constructed, and so he began to prepare a comparative vocabulary of the European languages. The first edition was published in 1909 [141]; his approach to the task was suggested in the two-page note 'Mensura de internationalitate' [139].

In the meantime, Esperanto had replaced Volapük as the leading artificial language, and Peano was becoming acquainted with various proponents of that language, especially the French mathematician Louis Couturat. Already at the Second International Congress of Mathematicians in Paris in 1900 a note of the mathematician Charles Méray, discussing Esperanto, was read, and later Méray gave Peano a copy of his article 'L'Esperanto, langue auxiliaire internationale' of 1 May, 1901, (published in *L'Enseignement Chrétien*.) Indeed, there is evidence that Peano was in contact with Méray earlier.

Like Volapük, Esperanto was the invention of one man, Lazarus Ludwik Zamenhof, but unlike Schleyer, Zamenhof did not claim 'inventor's rights'. In his hometown of Bialystok (in northeastern Poland) there were Russians, Poles, Germans, and Jews. Zamenhof claimed that the diversity of languages early led him to the idea of creating a simple language that could be used by all and that he had already created the elements of such a language in 1878, while still in high school (and before the publication of Volapük.) He first published his language in July 1887, under the pseudonym 'Dr. Esperanto' — hence the name Esperanto ('hopeful') for the language.

By 1892 there was an Espero Society in St. Petersburg. The language then spread to Germany, due mainly to the efforts of Leopold Einstein, who converted the Volapük Club in Nürnberg. Volapük was then declining rapidly. By 1903, the periodical *Esperantista* was being published in Turin. The first international Esperanto congress was held in Boulogne in 1905, and was organized by A. Michaux. At that congress Zamenhof praised Schleyer and others who worked for the success of the international language movement, and he decided to invite authors of other languages to the next congress, which was held in Geneva in 1906. Thus Peano attended that congress by express invitation, as did Bonto van Bijlevelt, a proponent of Neutral. (Bonto visited Peano in Cavoretto in 1908, and Peano returned the visit in Brussels in 1909.)

The following year, in October 1907, Peano attended the meeting of the *Délégation pour l'adoption d'une langue auxiliaire*. Although the *Délégation* agreed in principle on a modified Esperanto, Peano observed the difficulty the delegates had in reaching any agreement. This did not prevent him, however, from cooperating with Couturat in the preparation of the mathematical dictionary that Couturat had long projected. This was published in 1910, the principal entries being in Ido (the modified Esperanto that was the language of the *Délégation*) with equivalent terms in German, English, French, and Italian. In his preface, Couturat thanked Peano for reviewing the Italian translations, noting that "Prof. Peano's services to mathematical science and international language are well known."

By 1908 Peano was convinced that the solution to the problem lay in abandoning dogmatic propaganda. He believed that the new international language would evolve through complete liberty and scientific cooperation. As Peano expressed it some years later [224, p. 233]: "The interlingua problem is scientific, and authority has no value in science. Liberty is the necessary condition." He knew, however, that some sort of organization was necessary to make the idea effective. He could have founded a new academy,

ACADEMIA PRO INTERLINGUA 129

but since he had already published Latino sine flexione, it would appear that his academy was formed to promote that language. There was, however, no existing academy that shared his views.

It was at this point that Peano decided to try to make over the old Volapük Academy, now *Akademi internasional de lingu universal*. The time was ripe, since the five-year period of stability, voted earlier, would come to an end that year. He presented himself, at the same time, as a candidate for membership in the academy and for the directorship. His credentials (a short biography and a list of his publications) were published in circular No. 48 (No. 93 of the original numbering) of the Akademi internasional [139.1]. This was in May 1908. Thus, when Peano expressed the wish at the Philosophy Congress in Heidelberg in September, that an academy be founded to promote the idea of an international auxiliary language, he had in mind something other than what Couturat supposed (or, perhaps, pretended to think.)

Peano was unanimously elected member and director of the Akademi internasional on 26 December, 1908. Of the six men proposed for the office of director, only Peano agreed to accept election. Among those declining to be considered was the old director, M. A. F. Holmes, who said, "A new director will be able to introduce new life into the Academy." Peano immediately began the task of reconstructing it. On 1 May, 1909, he issued his first circular and proposed for adoption the following two regulations: (1) The Academy grants the director the right to publish a journal in the name of the Academy. (2) Every academician, in the circulars and the periodical of the Academy, may adopt that form of interlingua that he prefers. The first of these was adopted by the Academy unanimously; the second passed by a vote of seven for and two opposed (who may bitterly have recalled Peano's promise in his application for the directorship to publish the circular in Idiom Neutral "for continuity.") As a consequence, the first issue of the *Discussiones* appeared at the end of the summer [146]. Peano introduced it with the optimistic statement, "The problem of an international language is near a solution." He continued to use Latino sine flexione, but was careful to safeguard the second rule proposed by him. He wrote in 1909, in fact: "The form that I used is not the language of the Academy, for I use Latinisms and irregular forms, where it is convenient for me. I am writing for the members of the Academy, and not in the language of the Academy."

Nevertheless, he believed that further efforts would lead to a common language and he pushed forward his own project of publishing a comparative vocabulary. The first edition, with 87 pages [141], appeared in 1909. He considered this project, a scientific approach to the internationality of vocabulary, to be of prime importance, and a second, greatly enlarged (352 pages) edition appeared in 1915 [169].

By the end of 1909, Peano had managed to have his friends Padoa and Pagliero elected to membership in the Academy. He was now ready to complete the process of liberalization that he had begun with allowing each academician to use his own language. He now proposed the final step — that the membership in the Academy be open to all. This radical step was accompanied by a change of name, the old *Akademi internasional de lingu universal* disolving itself and giving all rights to the newly formed *Academia pro Interlingua*. The restructuring process was now complete, and Peano remained as director of the *Academia* until his death in 1932.

Peano's dedication to the international auxiliary language movement is reflected negatively in the smaller number of mathematical publications, the only one of the year 1909 being a short note discussing the notation of vector analysis [142].

On 14 May, 1909, one of Peano's earliest and most faithful followers in mathematical logic, his dear friend Giovanni Vailati, died. Peano wrote a brief necrology which was published three days later in the *Gazzetta del Popolo* (of Turin) and later reprinted (along with other commemorations) in the *Bollettino di Matematica* [142.1]. Since most of Vailati's publications were in the form of articles and reviews scattered through a large number of journals, it was felt that a fitting tribute would be the collection of his writings into one volume. This was done and published in 1911 as *Scritti di G. Vailati*, a volume of just over 1000 pages. The editors were Mario Calderoni, Umberto Ricci, and Giovanni Vacca. Peano, along with Gino Loria, Vito Volterra, and Orazio Premoli, helped prepare the bibliography, and Peano helped read page proofs. He also subscribed 60 lire for the book. (The announced prepublication price was 12 lire.)

Peano was in Lucerne, Switzerland, from 3 to 10 August. The following month he was in Padua for the second congress of Mathesis, 20—23 September. On the afternoon of 22 September, Padoa and Loria jointly proposed: (1) that professorships in the methodology of mathematics be established in the universities and (2) that practice teaching be required for a teaching certificate in the middle school. Peano expressed his complete agreement with them.

Immediately on his return to Turin, Peano wrote to Mittag-Leffler to enquire if he would be willing to publish in the *Acta Mathematica* an article on definitions in mathematics that Peano would write in Latino sine flexione. Apparently Mittag-Leffler would not agree to this; at any rate no article was published. (Peano may have met Mittag-Leffler in 1899 when he was in Turin, although Peano did not attend the 12 March session of the Academy of Sciences at which Mittag-Leffler was present. There are postcards from Peano, dated 27 July, 1899, and 13 February, 1901, thanking Mittag-Leffler for sending offprints of his work. A letter from Peano of 21 February, 1924, is apparently in reply to a request from Mittag-Leffler for information about Cauchy's contact with the Academy of Sciences of Turin. Peano describes the written documents of Cauchy at the Academy, noting that "the Royal Academy of Sciences of Turin never published anything of Cauchy."

At the 30 January, 1910, meeting of Mathesis he read a paper on the fundamentals of analysis [143]. In this paper we find Peano's mature views on teaching, and once more we find him calling for rigor and simplicity. "Mathematical rigor is very simple," he tells us. "It consists in affirming true statements and in not affirming what we know is not true. It does not consist in affirming every truth possible." Not only the principles he states, but even some of the examples he gives may still be read with profit by today's teachers.

Peano's mother died this year, having survived her husband 22 years. Her tombstone reads simply: "Peano Rosa, née Cavallo, model mother, deceased 23 May, 1910, at the age of 75."

Already in his calculus text of 1884 Peano had briefly discussed the orders of magnitude of functions approaching infinity. He returned to this topic in a short article presented to the *Lincei* Academy on 16 June, 1910, [144]. His point of departure was a text of Borel published that year, in which Peano finds fault with a couple of definitions. (He gives credit for noting this, however, to V. Mago, citing his unpublished degree thesis.) Peano then proceeds to define (by abstraction) a new set of objects which he calls 'fini'. These correspond to Borel's orders of magnitude, but Peano is able to define an arithmetic for these entities, whereas he had found fault with Borel's definition of multiplication. He comments:

We thus have new categories of the infinite and the infinitesimal that are actual, or constant. We see from this that the difference between the actual, or constant, infinite and the potential, or variable, infinite – a question that has deeply interested philosophers of mathematics, and continues to do so – is purely a matter of language.

In the meantime, Peano's teaching continued to decline in the eyes of the majority of his students and colleagues. Alessandro Terracini, who attended the lectures on Higher Analysis that year, recalled in his autobiography that Peano's lessons, like those of calculus, consisted of leafing through the

Formulario, stopping now and then on some point. All the same, he seems to have learned the material, for when he reported the result of the examination by D'Ovidio and Peano to his brother, he wrote a postcard in *terza rima*, concluding with the triplet:

> D'Ovidio then shook the tail of his coat,
> While Peano stroked his venerable beard
> And gave 'highest honors' the final vote.
> [Agitava D'Ovidio allor le code,
> E, del mento tirandosi l'onore,
> pieni voti me dié Pean con lode.]

But Peano was relieved of the course in Higher Analysis and it was entrusted in the fall to Guido Fubini, who had been a professor at the polytechnique since 1908.

Many of those who wrote theses under Peano's direction were women and, according to Terracini, they were not always well prepared, but the 1910 graduate, Maria Gramegna, was one of the most promising. In fact, Peano had presented a long article by her on differential and integral equations to the Academy of Sciences on 13 March, 1910. In it she anticipated the modern application of matrix theory to the study of systems of differential equations; the idea for this probably came from Peano. Her abilities were not to be realized, however, for she went to teach at the Normal School in Avezzano (L'Aquila) and died on 13 January, 1915, a victim of the earthquake that destroyed that town and killed 96% of its inhabitants.

In the spring P. E. B. Jourdain was preparing a long article on 'The development of theories of mathematical logic and the principles of mathematics.' Some of this article was published in 1910, but the section concerning Peano did not appear until 1912. Jourdain sent Bertrand Russell a draft copy of this section for his comments. Russell returned it on 15 April, finding nothing to criticize in it, but his brief comment, in the accompaning letter to Jourdain, is of some interest for its judgment of Peano's contributions and his character. He wrote:

I read carefully all the pages you specially asked me to look at. I think the introduction of ϵ was Peano's most important contribution, but the emphasis on implications with variables was also a great step. The reason I think it less important than ϵ is that the *practice* had always existed, e.g. in Euclid, and that Peano, tho' he emphasized such implications, did not give a very clear account of their nature.

I think the article should go to Peano. He is not touchy, so far as my experience goes, tho' he does not himself readily adopt improvements made by others.

Jourdain followed Russell's advice and sent the article to Peano, who also approved its publication and added a number of notes. Peano had by now given up active research in this field, but some of his remarks at the time are valuable for showing how his idea grew and changed. For example, he notes that in 1888 he began his exposition with the calculus of classes, but in 1889 with the calculus of propositions. In 1910 he prefers the calculus of classes, as being "more precise and rigorous." (He returned to this order in 1897.) He admits that several of the propositions in the first *Formulaire* were not precise, and even that the lack of a notation introduced in 1897 rendered the earlier theory of functions "confused and ambiguous." Perhaps as part answer to those who criticized him for not adopting the improvements of others (see Russell's comment above), he remarked: "On account of lack of time, the noteworthy and important theory of Russell could not hitherto be written in the *Formulaire*."

There is another mention of Peano's ϵ and its importance for Russell in his letter to Jourdain:

Until I got hold of Peano, it had never struck me that Symbolic Logic would be any use for the Principles of mathematics, because I knew the Boolean stuff and found it useless. It was Peano's ϵ, together with the discovery that relations could be fitted into his system, that led me to adopt symbolic logic.

Sometime in 1910, probably during the summer, Peano was in Berlin, where he was in contact with members of the international auxiliary language movement. His time was now mainly devoted to this, editorial work for the *Academia pro Interlingua Discussiones* [207] replacing his previous activity for the *Rivista di Matematica*. Vol. 1 of the *Discussiones*, began in 1909 and continued through 1910, with a total of 225 pages, articles written by Peano appearing on 61 of them [146–148]. He also managed to have a propaganda article printed this year in *Classici e neo-latine* [145].

In the beginning, Peano used the term 'interlingua' generically, but he soon began to limit its use to the language he saw evolving from the Academia pro Interlingua. Already in 1910 the Academy adopted 'rules for Interlingua.' There were three: (1) Interlingua uses every word common to English, French, Spanish, Italian, Portuguese, German, and Russian, as well as every Latin word with English derivatives. (2) Every word, which exists in Latin, has the form of its Latin root. (3) The suffix -s indicates a plural. These rules were, of course, only advisory and no one was obliged to follow them. They represented a majority opinion, however, and in fact most of the articles published in the *Discussiones* resembled Peano's Latino sine flexione.

To assist in selecting the proper form of a word Peano had already published a vocabulary in 1909. With the adoption of rule No. 2, he seems to have hoped that an ordinary Latin dictionary would be adequate for most writers of the language. He saw the value, however, of a more complete Interlingua dictionary and published a second, greatly enlarged edition in 1915. Indeed it is not clear just what was meant by rule No. 2, since Peano explains the 'root' of the verb *amare*, for example, as *ama*, but chooses the ablative form for most nouns. It is clear that rule No. 2 implied following the Latin spelling, if the word were of Latin origin, since otherwise the reader would find an ordinary Latin dictionary unsatisfactory. Any proposition subscribed to by six members was officially discussed and voted upon, although the academy voted surprisingly few restrictions on its members. Peano's own liberal attitude seemed to have generally prevailed. It should be noted in this connection that after opening membership in the Academy to anyone paying ten francs to the treasurer, Peano's friends Tommaso Boggio, Burali-Forti, and Giovanni Vacca joined the Academy early in 1910, as did D'Ovidio and Gustavo Sannia (D'Ovidio's nephew and assistant.) P. E. B. Jourdain was also persuaded to join that same year. (By 1927 there were 365 members, including 17 in the United States.)

By now Peano seemed to be devoting all of his energy to this project. Indeed, after receiving a copy of Whitehead and Russell's *Principia Mathematica*, he wrote to Russell on 16 December, 1910: "Thank you for the book *Principia Mathematica*, which I mean to read carefully. At present, the time I have free from school is occupied with the question of Interlingua, which I do not believe so absurd as the majority of people tend to believe. I am sending you some of my articles on this subject. Please thank Mr. Whitehead also."

If Russell showed this note to his new student Henry M. Sheffer, it may have influenced Sheffer's plans. When he was awarded a Sheldon Traveling Fellowship from Harvard University for the year 1910–11, Sheffer had proposed to spend the first half of the year in Italy because of Peano, and the second half in Germany (Cantor, Frege) and England (Whitehead, Russell). On 23 October, 1910, however, he wrote a 'report' to Dean Briggs, saying that he had been a month in London, two months in Paris, and was then taking two courses from Russell at Cambridge. He planned to go to Bologna during the Easter recess to attend the International Philosophical Congress and "remain for a while with Peano in Turin." Sheffer may have met Peano in Bologna, but it is unlikely that he went to Turin. Peano would undoubtedly have suggested that he go right back to Russell!

Sheffer may have been a source of information for Norbert Wiener, who won a traveling fellowship from Harvard in 1913. Wiener later wrote in his autobiography: "Two places suggested themselves as alternate destinations: Cambridge, where Russell was then at the peak of his powers, and Turin, which was famous for the name of Peano. I learned that Peano's best days were over, and that Cambridge was the most suitable for a training in mathematical logic" [Wiener 1953, 179].

CHAPTER 19

APOSTLE OF INTERLINGUA

In a note presented to the *Lincei* Academy on 8 January, 1911, Peano announced the publication of Volume 1 of Whitehead and Russell's *Principia Mathematica* [149]. Later, in a letter to Russell, he referred to this note as a 'review-notice'. It was, in fact, less a review than a mere announcement of publication, since Peano discusses only the definitions of 'relation' and 'function' contained in it, and this only to show that these may be written in the symbolism of the *Formulario*. He would elaborate this in a more formal review published in 1913 [163]. At the Academy of Sciences he was less active than usual, attending only three sessions during the academic year 1910–11, and then only to present papers written by others.

This year marked the 50th anniversary of the united Italy. Among other celebrations of this event were the inauguration on 5 June of the Victor Emanuel II monument in Rome and the international exposition in Turin, which opened on 20 April and closed 19 November. Because of this last attraction, Peano was probably able to greatly increase his personal contact with advocates of an international auxiliary language.

The one problem in mathematics that continued to occupy him was the question of definitions in mathematics. He published a long article on the topic this year in a Spanish journal [151], but did not come to any final conclusions. Apart from this and the announcement of *Principia Mathematica*, his few other publications this year dealt with Interlingua. There was a propaganda article in French, explaining his Latino sine flexione [150], a note in the *Discussiones* (20 July) concerning the passive verb form [152], and a separate sixteen-page pamphlet (21 October) giving 100 examples of Interlingua [153]. This last also included an Interlingua-Italian vocabulary. There was a second edition in 1913, with an Interlingua-Latin-Italian-French-English-German vocabulary.

Peano was in Bologna for the Fourth International Congress of Philosophy, 6–11 April, 1911. It is indicative of Peano's new dedication to Interlingua that whereas he originally intended to speak on mathematical logic (and was so listed in the program announcement in March), he actually presented a paper on 'A question of rational grammar,' in which he discussed his view of the arbitrariness of the grammar of natural languages [154]. Couturat was there, again talking about the connection between logic and linguistics in the problem of an international language. The Germans and English-speaking philosophers for the most part avoided this congress. On Tuesday, 11 April, Peano was co-chairman (with Eugenio d'Ors) of the session of logic and theory of science.

In September 1911, Peano was on a trip to Switzerland. With the opening of this academic year in Turin, Peano is 20th in seniority among the Ordinary Professors of the university. His schedule of classes remains the same: lectures on infinitesimal calculus at 10:30 Tuesday, Thursday, and Saturday, with applications and exercises at 8:15 Monday, Wednesday, and Friday. His final publication of 1911 was a brief note giving some etymological observations [154.1]. This was published in November in a journal of the Spanish Mathematical Society and is Peano's only publication originally in Spanish. In the 10 February, 1912, issue of the *Discussiones* he elaborated these ideas in a discussion of the derivation of one grammatical form of a word from another [155].

Taking advantage of the Easter recess at the university, Peano traveled south, arriving in Syracuse on 5 April. On 14 April his note 'On the definition of probability' was presented to the *Lincei* Academy [156]. In it he shows how the symbolism of the *Formulario* may be used to give a set theoretic definition of probability. He begins by accusing the usual definition of being a vicious circle, but it is difficult to see, in his note, how he believes he has overcome this difficulty. Nonetheless, his analysis displays his usual concern to simplify and place on a strictly mathematical basis the foundations of the subject.

One of the criticisms leveled against Peano by his colleagues was that he never tested his students, and even presented candidates for the final thesis examination who were totally unprepared mathematically. Partly as a result of one such painful experience, a comprehensive mathematical examination was established. This was not laziness on Peano's part or a lack of concern, but derived from a deeply felt principle. There is no need for us to torment such a student, he would say, "life will flunk him." Indeed, he carried this principle to its final conclusion and would abolish all school examinations. In an article in the popular periodical *Torino Nuova* of 17 August, 1912, entitled 'Against the exams,' he set forth these views publicly [158]. He deals primarily with the schools, saying: "It is truly a crime against humanity to torment the poor pupils with exams, to be assured that they know things that the general educated public does not," but he immediately adds: "This applies as well to the secondary schools and the university."

In a letter of 20 March to Bertrand Russell, Peano announced his intention to present a brief paper at the Fifth International Congress of Mathematicians, to be held in Cambrige, 22–28 August, 1912. The note he presented there, dealing with existential propositions [157], is not of great importance for mathematics, but no doubt Peano saw this as a golden opportunity to promote the international language by presenting it in Interlingua. As it turned out, he was not allowed to do so. The regulations allowed only the four languages (English, French, German, Italian) and in the end, despite a direct appeal to Russell for help, Peano was forced to present the paper in one of the official languages. He chose Italian. The episode has a tragicomic element, for in Peano's appeal to Russell he is willing to affirm that his Latino sine flexione *is* Italian, which it certainly is not, just for the chance to use it. "I believe," he wrote, "that the experience would be profitable." Ironically, by using Italian he was almost certainly understood by fewer mathematicians present than if he had used French (which he could easily have used, and rightly chose in 1900). Indeed, it is doubtful if many more would have understood him had he been allowed to speak in Latino sine flexione. It is small wonder that he returned from the congress sadly disappointed with the way the English pronounced his language.

But Peano's paper did give rise to a discussion that included Ernst Zermelo and G. Itelson, and was continued at the next session by E. H. Moore, with a reply by Peano. Nor did the episode prevent Russell from writing to Ottoline Morrell on the first day of the congress, "He looks the ideal picture of a logician; he has quite extraordinary nobility from singlemindedness – I have a very great reverence for him."

Having neglected the Academy of Sciences for a couple of years (except to present for publication the papers of his friends), Peano offered that body on 17 November, 1912, the result of his research into the relations between derivatives and differentials [159]. It was a masterly study, both from a historical and a critical viewpoint. In the 19th century, the mathematical community had struggled to come to terms with differentials, but in the process, according to Peano, the meaning attached to the terms and notations of Leibniz, Newton, and Lagrange had become distorted. Peano goes back to the sources and shows, in support of his thesis, that even the earlier notation was incorrectly quoted. His thesis is that these pioneers *thought* in derivatives, not differentials, as was commonly supposed. Indeed, Peano goes so far as to say that "wherever *differential* is written, one may read *derivative*, and the truth of the proposition remains." (This is almost an echo of the remark of Karl Marx that "$dy = f'(x)dx$ thus appears to us as another form of $dy/dx =$

$f'(x)$ and is always replaceable by the latter." In his analysis of the differential, however, Marx went on to stress its role as an operator.)

Superficially a study of the notation for derivatives and differentials, his article contains acute critical observations throughout, and is everywhere scholarly. The article concludes with a long passage from Poincaré (who had died the previous July.) Peano quotes approvingly, though he has a few critical comments. As usual, Peano avoids a polemical tone — just the oppsite of Poincaré, who had said that one ought to know German "although that language has ridiculous rules of construction and an alphabet which has no common sense about it," because it is spoken by many learned people.

At the 17 November session of the Academy of Sciences, Peano and Somigliana had been asked to report on a memoir by Sannia; Somigliana read their favorable report on 15 December [159.3].

In 1911 Castellano published a second edition (the first edition was in 1894) of his course in rational mechanics at the Military Academy. Peano wrote a review in 1912, recalling the pioneering efforts of Castellano [159.1]. In May of this year he wrote a preface for the publication in book form of Padoa's long article on deductive logic (Padoa's major publication) [159.2]. In it he says that Padoa was formerly his "distinguished student." This was not literally true, but Padoa always said that he was "proud of being a disciple" of Peano.

Johann Martin Schleyer, inventor of Volapük, died on 16 August, 1912, at the age of 81. Peano wrote a necrology for the *Discussiones* [159.4]. Schleyer had probably not visited Italy since 1875, when he climbed Mt. Vesuvius, and he almost certainly ignored the Academia pro Interlingua. Peano uses the occasion, of course, to trace the history of the Academia pro Interlingua as well as to tell of Schleyer's pioneering efforts. He claims the establishment of "peace and complete concord only in 1910, after the elimination of the dogmatic articles from the old statutes," and closes on the optimistic note:

That book of Schleyer remains an imperishable monument of the first form of Interlingua, nearly practicable; and it remains as a model for all successive forms. A system of interlingua, if it is to be successful, however, must be simpler than Volapük in every point. Schleyer's Volapük lives in every system of interlingua, as also in the definitive system, which is imminent.

Work for the progress of Interlingua continued into 1913 with the publication of Volume 4 of the *Discussiones*, but this was the last year of its publication. Peano contributed a note on the vocabulary of Interlingua to the 8 February issue [160], which was reprinted in an Academia pro Interlingua 'circular' of 1924, as well as in the first volume of *Schola et Vita* in 1926. (*Schola et Vita* was the most important organ to promote Interlingua.)

Peano was also becoming more involved with Mathesis, the organization of secondary school teachers of mathematics. At the 13 April meeting of the Piedmont Section, for example, there was a discussion of the merits of introducing mathematical logic into the middle school. One might have expected Peano to approve, but he did not. Rather, he is in favor of making sure the text used is rigorous, and by this he means that the definitions are exact and the proofs correct. He even says that the teacher may omit proofs that are difficult or give no insight to the pupil. What he is opposed to is the substitution of simpler, but false, proofs for complicated ones.

The same day (13 April, 1913) he presented a paper to the Academy of Sciences that was prompted by other discussions concerning the advantages of introducing limits and derivatives into the secondary school mathematics program. [161]. This paper was a critical analysis of the definition of limit found in school and university texts then in use, but it also ranged over other topics. For example, Peano has some acute remarks concerning Dedekind's 'creation' of real numbers. In contrast with his statement regarding school algebra, Peano feels that the introduction of mathematical logic would be useful *if* limits are to be introduced into the schools. His reasoning is based on the difficulties involved in discussing variables:

Ordinary language is an imperfect tool for stating propositions with several apparent variables. This becomes simpler with the distinction between universal and particular propositions, made by general logic, and better still with the equivalent symbols ∃ and ⊃ of mathematical logic.

(It should be noted here that Peano means the use of existential and universal quantifiers. The symbol ⊃ has the variable universally quantified as tacit subscript.)

A note by Paolina Quarra that Peano had presented to the Academy of Sciences on 30 March had included the calculation of the remainder terms in several quadrature formulas. That, and the work of several of his later students (up to 1924), was based on Peano's general method for calculating remainder terms of functions satisfying certain general conditions. Peano wrote an article discussing this method [162] and it was presented to the *Lincei* Academy on 4 May. Peano was perhaps the first to systematically derive quadrature formulas without using interpolation methods; his method was integration by parts. This method was later systematically used by R.

von Mises, who showed in 1935 how it is possible to derive every quadrature formula using only integration by parts.

Paolina Quarra graduated this year, along with twelve others. (For the years 1901–15, there was an average of seven graduates a year.) In the fall of 1914 she was listed in the university yearbook as 'voluntary assistant for infinitesimal calculus.' Pagliero was still Peano's regular assitant. (It is curious to note that the yearbook for 1913–14 lists a Cavoretto telephone number for Peano. It is 09.)

In 1913 Peano visited Venice, Vienna, Budapest, and Berne. He recalled the visit to Budapest in a letter of 26 July, 1929, to Dénes Szilágy, in which he says that the people knew no French and little Italian. They knew German, but would not speak it for political reasons. So he manufactured 'Hungarian sine flexione,' and with this he "spoke to waiters, policemen, and visited Marich." He was probably in Berne in late August to attend the Esperanto congress. (Bertrand Russell visited Turin in August and was disappointed at not finding Peano there.)

Peano wrote his last article dealing with mathematical logic this year. Appropriately it was a review of Whitehead and Russell's *Principia Mathematica* [163]. Two volumes had been published (and the third was in the press), but Peano's review deals only with Volume 1. He concludes with the remark: "Volume 2 treats finite and infinite cardinal numbers, operations on them, series, and limits of functions. I hope to be able to talk about it shortly." He never did, however. Naturally, he compares the work with the *Formulario*. After noting that the authors adopted many symbols of the *Formulario*, but introduced new symbols, he explains this by the different purposes of the two books:

In the *Formulario* mathematical logic is only a tool for expressing and treating proposi- tions of ordinary mathematics; it is not an end in itself. Mathematical logic is explained on sixteen pages; an hour of study suffices to know what is necessary for the applications of this new science to mathematics. The book of our authors treats mathematical logic as a science in itself, as well as its applications to the theory of transfinite numbers of various orders. This requires a much vaster symbolism.

Peano's first publication dealing with mathematical logic was in 1888. His publications on this topic thus cover 26 years and form what Peano himself called the "complete" *Formulario*. Peano's original contributions to the subject, however, were all in the first half of this period, ending, ironically, with his meeting with Bertrand Russell. Russell quickly saw this meeting as a turning point in his own intellectual life. Peano did not – but such, in some respects, it proved to be. He was the recognized leader in the field of mathematical logic in the decade before 1900; after 1900 Russell was the leader.

Peano's friend Mario Pieri died on 1 March, 1913, of cancer, after a long illness. Peano wrote a brief necrology for the *Discussiones* [163.1]. (Pieri was a member of the Academia pro Interlingua.) In it he discusses Pieri's contributions to mathematics and its teaching. He also praises the man:

Pieri was totally dedicated to science and teaching. He was an untiring worker, honest, and of a singular modesty. When, some twenty years ago, the professors in Italy agitated for higher salaries, Pieri declared that their salaries were already above the work they did and their merit.

The last issue of the Academia pro Interlingua *Discussiones* (30 October, 1913) contained a brief note by Peano dealing with questions of grammar [164]. At the Academy of Sciences on 30 November he read his and D'Ovidio's favorable report on the memoir of Vincenzo Mago, that had been presented by Peano the previous April [165].

In January 1914 the Belgian journal *Mathesis* published in Latino sine flexione another article by Peano dealing with the remainder term in quadrature formulas [166].

With the termination of the *Discussiones*, the *Revista Universale* became the official organ of the Academia pro Interlingua. Its editor was Ugo Basso in Ventimiglia (a coastal town near the frontier with France.) To become a member of the Academia pro Interlingua, one had only to send ten francs to the *Revista Universale*. Only a very limited number, however, were named 'Professors of Interlingua'. They were, at this time: Ugo Basso (Ventimiglia), J. B. Pinth (Luxembourg), J. Meysmans (Brussels), G. Moore (London), G. Pagliero (Turin), G. Peano (Turin), and J. Bernhaupt (Beirut). (Pinth and Bernhaupt had been the last members of Schleyer's academy, but had left it in 1908.)

Peano wrote two necrologies for the *Revista Universale* this year. The first was of Augusto Actis and appeared in the March issue [167.1]. Actis was one of the original members of the Volapük Academy, founded in 1887. He was born on 27 October, 1831, and was a professor of Latin and Greek from an early age. He was named 'emeritus' by the Academy in 1912, having approved of the various modifications that led to the Academia pro Interlingua. He died in Bologna on 5 January, 1914.

Peano's tolerance for other forms of interlingua than his own Latino sine flexione is pointed up by his publication this year (on 1 August) of a sixteen-page pamphlet (written in Latino sine flexione) explaining the fundamentals of Esperanto [167]. This is probably one of the few books or pamphlets ever written in one artificial language to explain another.

CHAPTER 20

THE WAR YEARS

In the summer of 1914, events were happening that quickly led to a general
conflagration in Europe. On 28 June, Archduke Francis Ferdinand of Austria
and his wife were assassinated at Sarajevo. One month later Austria-Hungary
declared war on Serbia. On 1 August, Germany declared war on Russia,
France began mobilization, and Italy declared its neutrality. That same day
Peano went to Paris to see for himself what the war situation was, taking a
young relative with him. He found out soon enough, when Germany declared
war on France on 3 August and invaded Belgium. On their return to Italy,
Peano became separated from his companion (probably his nephew), who,
because Peano had his passport, climbed to the top of a railway carriage in
order to cross the border. Peano later congratulated him on having escaped,
apparently undisturbed at having caused the problem.

One of the earliest casualties of the war was Louis Couturat, who, the very
day Germany declared war on France, was riding in a carriage that was struck
by the automobile carrying the French orders for mobilization. Couturat was
a noted pacifist, but Peano was probably unaware of the irony of the event
and, in his necrology of Couturat (published in the *Revista Universale* in
October [167.2]), merely mentions that he died on 20 August as a result of
an automobile accident. Peano praises Couturat's scientific researches into the
international language movement (which "raised the study of interlinguistics
to a science"), but he criticizes Couturat's spirit of partisanship for Ido,
saying that his dogmatic propagandizing caused the number of its adherents
and his friends to diminish. (But Peano was probably unaware of the leading
role Couturat played in the intrigue through which Ido was selected as the
language of the *Délégation pour l'adoption d'une langue auxiliaire*.)

At the Academy of Sciences on 27 December, Peano read his and Segre's
favorable report on a memoir of Burali-Forti [168], which the secretary had
presented, for Peano, on 13 December.

In the meantime, Peano was busy with his own scientific research in
interlinguistics, and was seeing through the press the second, greatly enlarged,
edition of his comparative vocabulary [169]. It is dated 1915, but was
probably completed in December 1914. He gave a copy to the Academy of
Sciences on 10 January, 1915. The title is *Vocabulario commune ad latino-*

italiano-français-english-deutsch, but the Greek, Spanish, Portuguese, Russian, or Sanscrit, and even Indo-European are also given for many of the some 14,000 entries. Cassina believed that this work alone would have been sufficient to make Peano famous. Certainly he was very proud to it, perhaps next only to the *Formulario Mathematico*. He gave a copy to his brother, inscribing in it: "To my dearest brother Bartolomeo, a souvenir of Giuseppe."

At the Academy of Sciences, Peano presented on 21 February another note dealing with the remainder term in quadrature formulas [170]. This time he treats Simpson's Rule, preferring to call it the Cavalieri-Simpson Formula. ("Not wishing to go too much against usage," he says, "I allow myself to attach the name of its first author.") A method somewhat different from that used in the 1913 article presented to the *Lincei* Academy is used.

In the meantime he determined to assume a more active role in the 'continuing education' of secondary school teachers. He, Boggio, and Bottasso decided to sponsor a series of mathematical conferences. It seems clear that he did not intend this as a rival to Mathesis, but rather as complementary to it. An organizational meeting attended by about 40 people was held on 27 February at which Peano explained that the purpose of the conferences was to serve as a means of exchanging ideas on elementary mathematics. He then recognized the value of Mathesis, displayed five journals that had articles on elementary mathematics, and invited those present to write about the same subjects. Boggio next gave a talk. Finally Boggio was elected president of the conferences and it was announced that Bottasso would give a conference on 6 March. These conferences were held on the Saturdays of the academic year and continued throughout the war. In the university yearbook for 1919–20, Peano listed 26 publications, other than his own, that resulted from these conferences [192]. Many of these were presented by Peano for publication in the *Acts* of the Academy of Sciences.

As evidence of his concern for even small details in the teaching of mathematics, Peano published, in the March issue of one of the teachers' journals, a brief note giving his opinion on the question of writing products of quantities [171], whether, for example, pupils should be taught to write 2 m $\times$ 3 m = 6 m^2 (where m stands for 'meters'). He is in favor of this usage.

About this time Peano printed a list of his publications (the last included is [171]) and he added the comment: "My works refer especially to infinitesimal calculus; and they have not been entirely useless, since according to competent judgment they contributed to the current constitution of this science. I never asked for extra courses to teach, and so I have been little disturbed in my studies."

His next publication was in the April issue of the journal of Mathesis and is a critical-historical study of Euclid's definition of irrational numbers [172]. This article was written in Latino sine flexione and was part of his effort to spread the use of this language.

At the Academy of Sciences on 25 April, D'Ovidio read his and Peano's favorable report on a memoir by Sannia [174]. On 13 June, Peano presented a paper [175], in which he again considers Cauchy's *grandeurs coexistantes* that he had discussed in his *Applicazioni Geometriche* of 1887. At the meeting of the combined sections of the Academy which closed the academic year on 20 June, the president gave a speech urging all to work for a greater Italy. The Academy applauded vigorously and the session ended with cries of "Long live Italy! Long live the king!" Italy was now at war.

The events leading to a declaration of war included demonstrations throughout Italy — and of course university life was disrupted. The university was closed by ministerial decree on 22 May, the day before war was declared, but disorder prevailed throughout May as the students impatiently followed the political events. Among other things, they set the university bell continuously ringing. When the announcement of war came the evening of 23 May, the rector of the university saw the students parade, "waving a flag and kissing it with tearful eyes, crying with the joy of finally being able to join the army to rescue our lands from the artillery of the two-headed eagle."

The first examination period had been set for 24 May to 24 June, but few asked to be examined. Many joined the army immediately, and many more registered for the accelerated officer training course. At the opening of the academic year in the fall, the rector gave a speech, honoring those students who had already fallen in battle. The list included one mathematics student, Alessandro Mazzantini, from Argentina.

Peano, too, saw the war as patriotic and Italy's side as right. Like most others, he hoped that it would not last long and that a lasting peace would be established. In anticipation, he had taken steps to insure the continuity of the Academia pro Interlingua. Elections for officers of the Academy had been scheduled for January 1915, but international communication was becoming difficult, so in a circular of 29 December, 1914, he proposed that the Academy confirm in office its current administration until January 1916. Only 26 members replied, but they were all favorable. Peano reported this in a circular of 2 April, 1915, adding: "Therefore we shall keep our society alive until next year, when, we hope, to this horrible war will succeed peace and a better order which will prevent violence and a future war." Like many another, Peano's estimate of the length of the war was far too optimistic and it was not until 1923 that he would call for a revival of the Academia pro Interlingua.

Peano was not, however, a super-patriot and, while condemning the Austrians (or rather their leaders) for starting the war, he found one of the causes in the diversity of languages. One of his most 'political' writings was published in the spring of 1915 in an English Journal edited by G. A. Moore, *The International Language; a journal of 'interlinguistics'*. Titled 'War and language' [179], it begins:

> That emperor who in August 1914 set Europe on fire, who was the first to order his army to invade neighboring territories, and condemned millions of men to death, has committed a crime more colossal than any other in the history of mankind. But material for the conflagration was prepared a long time ago, and by many men.
> One cause of war is the variety of language. . . .

Patriotism, he says, is "a collective pride that induces us to judge other people inferior," and he calls this opinion "erroneous, dangerous, and damaging."

The article closes with an anticipation of the League of Nations: "Let us hope that this war will instruct the public on the origins of the present evils, and of the necessity of a confederation of states throughout the world, which will permit the suppression of every army, and the transformation of the instruments of war into the tools of work." He elaborated this idea the next year in an article published in Turin, in an evening newspaper, on 8 March, 1916, with the title, 'The United States of the World' [m]. The idea was not new, of course, and indeed Peano's point of departure is a book with the same title written in 1914 by the Swiss A. Forel. In his article Peano urges that the Italian-French-English-Russian alliance transform itself into a confederation with one parliament and one army, the better to prosecute the war, and actualize the principles of liberty and equality.

Italy entered the war on 23 May, 1915. Shortly before, in April and May, Peano made several trips, going no further north, of course, than Switzerland. This spring he visited Brescia, Verona, Venice, Pordenone, Udine, and Lugano.

With international communication disrupted, the activity of the Academia pro Interlingua ceased and, after 1915, he published no article in Latino sine flexione until 1922, when he began formal publication of the Academia pro Interlingua *Circulare*. His last publication of 1915 in Latino sine flexione was a minor note in an even more insignificant journal. The note was on international prepositions, and the journal was *World-Speech*, published in Marietta, Ohio (this issue dated November 1915) [173]. Peano was always looking for propaganda outlets, but it is doubtful if he ever saw a copy of

World-Speech. The journal promoted *Ro*, the language invented by Edward
P. Foster, an ex-minister, who seemed to feel that only Ro would save the
world.

Of somewhat more moment was an article on the importance of symbols
in mathematics that Peano published in September [176]. In it he defends
mathematical logic, while once again affirming his own limited interest in it:
"Mathematical logic, which is useful in mathematical reasoning (and I made
use of it in this sense alone), is also of interest to philosophy."

Peano had introduced the term 'definition by abstraction' in 1894. Russell
had rejected definitions by abstraction in his *Principles of Mathematics*
in 1903, but Peano — and others — continued to defend them, and they
continued to be used by the majority of mathematicians. Peano discussed
various types of definitions by abstraction in an article for teachers published
in December 1915 [177]. The article concludes with a critical bibliography
of the subject.

At the Academy of Sciences on 26 December, Peano presented an article
dealing with the problems of printing mathematical formulas [178]. Here was
a question on which he surely was an expert! His principal recommendation is
that formulas be written on a single line, where possible. This, and other
recommendations, remain pertinent today, as authors of mathematical papers
continue to believe that formulas must be printed exactly as worked out in
their manuscripts. Peano's long experience in the matter shows very clearly
in such remarks as, "Letters which have to be cut, so as to put them as indices
on indices, become fragile and fall off in the printing." A glance at, say, *A
manual for authors of mathematical papers* of the American Mathematical
Society (1970) shows that editors must still ask for the changes suggested by
Peano.

Before the end of the year, he sent to the *Lincei* Academy a paper on
numerical approximations, which was presented on 2 January, 1916 [180].
He had begun studying this topic earlier and had directed the research of
several of his students and teacher-friends (and had presented their results to
the Academy of Sciences for publication.) This was the first of several papers
Peano published on this topic to which he gave much of his mathematical
activity in his later years. This paper treats rounding off errors in elementary
computations, and also derives several elementary differentiation formulas by
a simple application of increments.

In his note on definitions by abstractions in 1915 [177], Peano had noted
the resemblance of one of Leibniz' rules to one of the various statements of
the principle of identity in a logic text of 1910. He then remarked: "It is

probable that some scholar will be able to find a statement similar to this in books preceding Leibniz." In a note in the April 1916 issue of the same journal [181], he reported that his friend Giovanni Vacca had indeed found essentially the same statement in the writings of St. Thomas Aquinas and Aristotle. Peano gives the references and reports the passages. He reproduces these again in his entry on 'equality' for a 'dictionary of useful notions,' to which he also contributed entries on 'infinity', 'mathematical logic', and 'vectors' [184]. The articles on 'equality' and 'vectors' were later reprinted in the bulletin of Mathesis.

At the Academy of Sciences in 1916, Peano continued his practice of presenting for publication the work of his friends, especially those members of the 'mathematical conferences' at the university. The Academy was beginning to feel wartime shortages, however, and at a session of the combined sections on 21 May the president announced the suspension of the publication of the *Memoirs* and urged members to "be more severe" in accepting works of non-members for the *Acts*. During the following academic year, Peano did indeed present fewer papers of his friends.

A second study of rounding-off errors in numerical computations was presented to the Academy of Sciences on 14 January, 1917 [182]. In it, Peano presents a method of studying the complications that can arise as a result of rounding off, and gives several concrete examples. At the same meeting he took part in a discussion of ways in which the Academy could aid the government. This discussion was continued at the next meeting on 28 January, and Peano again took part. This was, of course, wartime, and its effects were being strongly felt.

At the sessions of 25 February and 11 March, Peano presented two parts of an article on numerical approximations [183]. In it, he follows up a suggestion in his January note and shows that the theory can be simplified and systematically studied by considering numerical approximations as intervals of numbers. He then develops general algorithms for operating with approximations. He calls these *graduale*, i.e. by degrees, and finds the origins of his method in the rules for multiplication introduced around 1600 by Kepler and others.

One of the ways of 'assisting the nation' suggested at the Academy of Sciences was a proposal of Nicodemo Jadanza, Professor of Geodesy, that a table of logarithms be prepared. Peano decided to ask the members of the Mathematical Conferences to lend their support to such a project, and called a meeting for 28 April to discuss it. D'Ovidio also attended this meeting. (Indeed, he was asked by Peano to preside.) Berardo Sterponi presented the case for an Italian table of logarithms, saying that the best Italian edition available was only a five-place table, and that imports were very expensive. As a result of the discussion, an 'order of the day' was approved calling for a commission to study the question of publishing a work "that would relieve the nation of the abuse of foreign booksellers, considering that it is not dignified for Italy to have to continually turn to outsiders for tables of logarithms." Peano, along with Sterponi and three others, was appointed to the commission. Sterponi read their report at the 26 May meeting of the Mathematical Conferences, that called on the state to publish a National Edition of a Table of Logarithms.

In 1895, Peano had been honored by the government by being made a Knight of the Crown of Italy. On 31 July, 1917, he was further honored by being promoted to Officer of the Crown of Italy — perhaps in recognition of his role in organizing and promoting the Mathematical Conferences of Turin. The conferences continued all during the war and that fall, at the first meeting of the academic year, Peano said:

The present hour in which our valiant soldiers are fighting for the defense of the country and for liberty might seem hardly opportune to begin the fourth year of the Mathematical Conferences. These are, however, most useful, especially in the present circumstances, since, in the name of mathematics, they serve to bring together the teachers scattered among the various institutes. Each of us, after having attended to our scholastic and civic duties during the week, finds here colleagues and, conversing of things mathematical, renews old friendships and makes new ones. This mutual instruction extends the frontiers of our knowledge and makes us better fitted for our mission.

Then, after describing several of the conferences, he goes on to say:

The Minister of Public Instruction, informed by the Rector of the University of these Mathematical Conferences, this very year, in a letter of 17 July 1917, wrote that he "was very pleased, expressed his thanks to the lecturers and their audience."

Given Peano's interest — and that of other members of the Mathematical Conferences — in numerical calculation, it is not surprising that a local publisher was persuaded to publish separately the tables in an engineering manual and to sell them at cost for use in the schools. This was done early in 1918, and Peano wrote a preface for it [185]. This small pamphlet was immensely successful, and in 1963 saw its eleventh reprinting. Peano mentioned this publication in a paper presented to the Academy of Sciences on 28 April [186], in which he discussed the problem of interpolating in mathematical tables. Extracts from this paper were included in later editions of the tables, to aid in their use.

Peano's former professor and long time friend Enrico D'Ovidio retired from university teaching this year. (He was 75 years old on 11 August, 1918.) In honor of the occasion his colleagues prepared a volume of various articles. Peano contributed a brief note on the remainder in interpolation formulas [187].

Matteo Bottasso died on 4 October, 1918. He had been one of the strongest supporters of the Mathematical Conferences until 1916, when he gave up his position at the Military Academy to become a professor at the University of Messina. He was only 40 years old when he died leaving a widow (he had been married one year) and a child. T. Boggio and Peano collaborated in writing his necrology [188].

On 11 November, 1918, the armistice was signed in the forest of Compiègne. The Italian troops would return to their homes in March. In the meantime, since October, all of Italy was suffering from an invasion of influenza from Spain. Peano, with returning optimism opened the fifth year of the Mathematical Conferences on 7 December, 1918, saying:

We are beginning our fifth year of these conferences, cheered by peace and the victory of liberty over violence, thanks to our valiant soldiers and allies, and to the high ideas of justice personified by our colleague, Professor Wilson. Let us hope that these ideas will fully triumph and that, the league of nations having been formed, all may work in the future for peace and progress.

Italy was not satisfied, however, with the aftermath of the war, and President Wilson's triumphal visit of the first week of January could not be repeated.

On 24 January, 1919, there died another of Peano's long time friends, Filiberto Castellano. He was one of the founders of the *Rivista di Matematica* in 1891 and a professor at the Military Academy until his death at the age of 58. Like Peano, he had a villa in Cavoretto and took an interest in the affairs of the local school. In his necrology [190], Peano calls attention to his use of vectors in his text in rational mechanics for the Military Academy. Castellano was also an active collaborator in the Mathematical Conferences, where he sometimes drew morals from mathematics. In a conference on 18 March, 1916, in the midst of the war, he discussed the logarithmic spiral and compared the progress of civilization to it:

Just as it is impossible to follow its curve into the past so as to find its origin, so also is it certain that this evolution will continue, despite the arrest apparent in crises such as the current stormy one, and will unfold ever more widely in its enlightened path.

CHAPTER 21

THE POSTWAR YEARS

Already before the war, the students under Peano's direction were mostly women and this remained so after the war ended and the soldiers returned. This was no doubt explained in great part by Peano's involvment with Mathesis and the Mathematical Conferences, where the majority of teachers attending were women. Many of the papers presented by Peano to the Academy of Sciences were by women, although in the academic year 1918–19 these included also notes written by soldiers still in service.

Peano contributed an article to the first volume of a journal of the mathematics of finance. (The issue containing it appeared in March 1919.) The editor had probably requested an article from the eminent professor expecting an article dealing with actuarial mathematics, since Peano had earlier worked in this field. That was some time ago, however, and the article that Peano actually wrote dealt with the signs used in algebra [189]. It is an echo of the paper presented to the Academy of Sciences in 1915, treating the difficulties in the setting of type for mathematical formulas. In the present article he also mentions the historical introduction of the various symbols discussed.

At the 27 April, 1919, session of the Academy of Sciences, one of the members asked that someone look outside of Turin for a cheaper printer for for the *Acts*. Peano, with his experience as an editor, and others, spoke up against this, pointing out the difficulties involved. We see from this, however, that the Academy was having financial difficulties. These difficulties would become acute in the following year and reach the crisis stage in the fall of 1920.

On 11 May, 1919, Peano presented a note to the Academy of Sciences dealing with the numerical solution of equations [191]. In it, he calls attention to the fact that the theory of numerical approximations had been studied by various members of the Mathematical Conferences.

The Mathesis Society held a national congress in Trieste, 17–19 October, 1919, attended by about sixty people. The president of Mathesis, Federigo Enriques, recognized Peano's leading role by proposing that he preside over the morning session of 18 October. At the close of the afternoon session there were cries of "Long live Trieste! Long live Fiume! Long live Italy!" reflecting the tense political situation in nearby Fiume (occupied by D'Annunzio since 8 September.)

With the opening of the academic year in the fall of 1919, Peano had become tenth in seniority among the Ordinary Professors. His assistant was still Pagliero, but his former student, Teresa Dusi, was listed in the university yearbook as 'voluntary assistant' (a 'position' held for a time by Paolina Quarra.) This would be Pagliero's last year as assistant. After him, Peano had only women assistants. In the yearbook, Peano also published a list of some of the speakers at the Mathematical Conferences, along with the titles of some of their publications [192]; most of these were women. By now the Mathematical Conferences had lost all formal structure. Here is how Peano described it at the first session of the academic year on 20 November, 1919:

The organization of these conferences is simple: With the permission of the Rector, the mathematics teachers may gather in the university halls every Saturday of the academic calendar at 5 o'clock. Each says what he wishes – there is neither president nor treasurer. Of the most important conferences, the author himself may take care of its publication, either in academic publications or in bulletins for teachers, according to the the nature of the subject.

In March 1920 Peano took an extended trip through southern Italy, visiting Rimini, Ancona, Foggia, Bari, Brindisi, Taranto, Cotrone, Reggio Calabria, Messina, Paola, Naples, and Rome.

Meanwhile, the Academy of Sciences was having financial difficulties, and on 11 April, 1920, Peano took part in a discussion of the mounting cost of publishing the *Acts*. It was eventually decided not to accept papers for the *Acts* during November and December of the next academic year, and the treasurer was asked to try to find a printer who would give better terms than Bona, who was then printing the *Acts*. This may partly explain Peano's complete lack of publications for the year, his first in 35 years. We cannot be surprised, however, that he was slowing in his mathematical production; he was 62 years old and his nearly 200 publications was already an enviable record.

In the fall of 1920 Elisa Viglezio became Peano's assistant at the university, a position she would hold for five years.

Having decided not to accept papers for publication in its *Acts* during November and December, the Academy of Sciences held no sessions for that purpose during that period, but other activities continued. A committee of three was appointed to decide the most important date in the life of Giovanni Plana, so that the Academy could plan an appropriate commemoration of its illustrious member. Peano was a member of the committee and he reported on 13 February, 1921, that the publication of Plana's *Théorie du mouvement de la lune* in 1832 was most significant. (A plaque was placed on the building of the Academy recalling that Plana completed his theory there.) The Academy had fallen on bad financial times, but the president was able to announce on 27 February that, as a result of the efforts of one of the members the military authorities had paid 5200 lire in compensation for damages caused the astronomical observatory during its occupation by them.

The financial plight of the Academy was further relieved by a gift of 20,000 lire from D'Ovidio's son-in-law Federico Petiva. This was announced at the 24 April session of another section and at the 1 May session of the natural sciences section of the Academy. It was also decided to let the newspapers know "so that not only may this act be properly recognized and appreciated, but also so that Cavaliere Petiva's example may find imitators."

In May, Peano published his last systematic study of definitions in mathematics [193]. His paper at the Congress of Philosophy, Paris, 1900, had dealt with the same topic, and some of those ideas are repeated, but he covers a good deal more ground here, discussing first the form of definitions in the writings of Euclid and Aristotle. He then discusses the relation of definitions to existence, 'possible' definitions, and the role of definitions by induction (recursive definitions) and by abstraction. Finally he discusses the role of primitive propositions and some of the difficulties in forming definitions. He concludes with his typical drive for simplicity: "Some mathematical definitions that are found in school texts are illusory. Just by leaving them out we gain in rigor and simplicity."

This publication was followed by Peano's first publication in Latino sine flexione since 1915. It is a brief note with a suggestion for defining the area of a rectangle in elementary textbooks [194]. A footnote reminds the reader that it is written in Latino sine flexione and calls attention to his *Vocabulario* of 1915. This article shows that Peano had not lost interest in Interlingua and marks the resurgence of optimism in its future. This renewed interest was also shared by others, as shown by Peano's being named to a committee of three to investigate Esperanto by the Italian Society for the Progress of Science, meeting in congress in Trieste. He was also consulted by the Committee on International Auxiliary Language in Washington and on 16 October, 1921, he furnished notes on the (unpublished) 'Handbook of Interlingua-English Draft.'

Peano's third (and last) publication of 1921 was a review of a calculus text written by his friend and colleague at the university, Tommaso Boggio [195].

Peano has high praise for it, especially for its use of the method of vectors, which he says "constitutes that royal road sought in vain since the time of Euclid."

On 15 October, 1921, Peano was again honored by the government by being promoted to *Commendatore* of the Crown of Italy.

At the first session of the academic year on 20 November, 1921, the Administrative Council of the Academy of Sciences announced a limitation on the number of pages that could be submitted for publication in its *Acts*. Each member was allowed 40 pages, of which no more than 20 could be for non-members, and this last could include no more than two notes. The regulation was probably not aimed precisely at Peano, although he was one of those most affected by it. He easily managed to fill his quota, with notes by U. Cassina on 22 January, 1922, and Burali-Forti on 19 February (taking 20 pages of the *Acts*), and one note of his own on 19 March (taking exactly another 20 pages.)

On 1 January, 1922, Peano sent out a circular from the Academia pro Interlingua, in an effort to revive that organization. The circular was not, however, mere propaganda, but also contained a brief article of 'Calculations with the calendar' [196] that was reprinted on 31 January in a scientific bulletin.

The paper that Peano presented to the Academy of Sciences on 19 March [197] (and published in translation into Latino sine flexione in August) was an elaboration of an idea in the brief note of the previous year, treating the area of a rectangle. In it Peano finds fault with the treatment of quantities in school textbooks. He justifies the physicists method of operating directly with quantities instead of the 'pure' mathematician's treatment of operating only with numbers. That is, he prefers to multiply length by width and get area, rather than say that area is the product of the measure of the length by the measure of the width, thus once more justifying the notation (3 meters) $\times$ (5 meters) = 15 (meters)^2.

On 4 June he published another Academia pro Interlingua circular, in which he reviewed A. L. Guérard's *A Short History of the International Language Movement* [198]. The next circular, issued 10 September, contained his statement of the 'Rules for Interlingua' [199], that had been adopted by the Academia pro Interlingua in 1910.

The fourth circular, dated 15 December, 1922, contained a note on the prospects of an international language for the League of Nations [200]. His other publications of 1922 included a note on logarithms, written in Latino sine flexione and published in the Polish journal *Wiadomości Matematyczne*

[201]. (The editor of this journal was Samuel Dickstein, an advocate of the international language movement and a member of the Academia pro Interlingua.) This note is a summary of a larger work by A. Borio, a teacher in the lyceum in Cuneo.

Also in 1922 there were two notes to a journal of elementary mathematics. The first [201.1] referred to two other notes: One had said that the binomial formula was inscribed on Newton's tomb. Peano remarked that he had visited the tomb (presumably in 1912) and that the formula was not there. He supported his claim with a photograph of the tomb and a copy of the inscription. The other note had told a mathematical riddle dealing with camels owned by Turks. Peano remarked that the story, as told by Leonardo of Pisa, involved not a Turk, but an Arab. "The Arabs," he wrote, "have taught us much, the Turks nothing."

In his other note to the journal [201.2], he gives several "practical" problems in arithmetic for schools. Some of these deal with the calendar, and he offers to send copies of his longer article on the calendar to the first fifty people requesting it.

In the fall of 1922 there began a reorganization of the mathematics faculty of the University of Turin. Previously the chair of 'algebraic analysis' had also included analytic geometry. This was the chair held by D'Ovidio until his retirement in 1918, when these subjects were entrusted to Boggio, so that, that previous year, 1921–22, the university yearbook listed Boggio as director of the School of Algebra and Analytic Geometry, with Ugo Cassina as assistant. According to the yearbook for 1922–23, Peano was Director of the School of Infinitesimal Calculus and Algebraic Analysis, with two assistants: Elisa Viglezio and Ugo Cassina. According to the published schedule, algebra was taught to first-year students and calculus to second-year students. The hours do not conflict, but most probably Cassina taught the algebra course.

The arrangement, however, was only temporary, and in 1923 there was established a separate School of Algebraic Analysis, with Ugo Cassina as director, assisted by Amalia Guglielmi. Cassina was also assistant to the School of Analytic Geometry, of which Alessandro Terracini was named director. This year, also, there was established a Chair of Complementary Mathematics, in conformity with the reforms of the Minister of Public Instruction, Giovanni Gentile (who had been appointed by Mussolini in October 1922.) This material was taught the first year by E. G. Togliatti and in 1924–25 by Boggio (who also taught algebra and analytic geometry that same year!)

A profound change took place in 1925 with the arrival of F. G. Tricomi, who was called from Florence to assume the chair of Algebraic Analysis. Tricomi has recalled, in his autobiography, the unexpected event that occurred at the first meeting of the mathematics faculty: When Peano learned that Tricomi was also to be given the chair of Complementary Mathematics, Peano approached him with the suggestion that they exchange teaching assignments, i.e., while Peano would nominally retain the chair of Infinitesimal Calculus and Tricomi that of Complementary Mathematics, that each teach the other's subject.

Now, Peano had been professor of calculus for 35 years, and while his teaching was excellent in the beginning, there is no doubt that it had degenerated to the point where efforts were made to get him to agree to some sort of compromise. He had, however, resisted any such suggestions. But now he spontaneously offered the faculty a way out! Tricomi immediately agreed (on the condition, of course, that the rest of the faculty approve — which they did.) This situation then lasted for six years, until in 1931, Peano officially transferred to the chair of Complementary Mathematics at the request of the rector, Silvio Pivano. There is no question that Complementary Mathematics better suited Peano's interests at the time, especially since it also included the training of high school teachers.

In 1923 Peano published a brief note in the *Academia pro Interlingua Circulare* [202.3] and three book reviews: of a 1919 Ido-German and a 1922 German-Ido dictionary [202.1], of an issue of a journal promoting Ido (the modified form of Esperanto espoused by L. Couturat) [202.2], and of a volume by A. Natucci on the concept of number [202].

In April 1923, Peano attended various meetings in Catania, Naples, and Rome, and he continued to take an active part in Mathesis, discussing at the 5 June session of the Turin section, for example, the effects of the Gentile reforms on secondary education. He also attended the national congress of Mathesis held in Livorno, 25—27 September, where he advocated further reforms, such as, that prospective teachers graduating in mathematics be allowed to obtain a degree in physics also with only one more year of study, and that those completing the scientific *liceo* be allowed to enter the university to study other subjects in addition to science and medicine. In general, Peano's position could be characterized as wishing to leave paths open to more opportunities.

Several times Peano had recommended that the teaching of arithmetic in the schools could be made more interesting by the introduction of more interesting problems and games, and he had given several examples. In 1924 he published a 64-page book (dated 7 March) containing just such material

[203]. (This book was several times reprinted.) The book is addressed to teachers, and its concluding paragraph is worth quoting:

The difference between us and the pupils entrusted to our care lies only in this, that we have traveled a longer tract of the parabola of life. If the pupils do not understand, the fault is with the teacher who does not know how to explain. Nor does it do any good to load the responsibility onto the lower schools. We must take the pupils as they are, and recall what they have forgotten, or studied under a different nomenclature. If the teacher torments his students, and instead of winning their love, he arouses hatred against himself and the science he teaches, not only will his teaching be negative, but the necessity of living with so many little enemies will be a continual torment for him. Each makes his own fortune, good or bad. Who causes his own troubles, cries alone. So said Jove, as Homer reported (Odyssey, I, 34). With these principles, dear reader and colleague, may you live happily.

CHAPTER 22

THE TORONTO CONGRESS

Peano had been a principal speaker at the First International Congress of Mathematicians, in Zurich, 1897. At the 1912 congress in Cambridge, he wished to present a paper in Latino sine flexione, but was prevented from doing so. The next congress was scheduled for 1916 in Stockholm, but could not take place because of the war. After the war, a congress was held in Strasbourg in 1920 that was, however, not entirely international and Peano did not attend. The next congress was held in Toronto, 11–16 August, 1924. According to the editor of the *Rassegna di matematica e fisica* [5 (1924–25), 18], Peano was "insistently invited by the Canadian government" to attend. This time he went and read his paper in Latino sine flexione. (He probably received assurances that this would be allowed before leaving Italy!) This was the highlight of the year, but there were other events in the spring.

On 23 February he gave a talk to the Mathematical Conferences in which he discussed the report of the committee of the Ministry of Public Instruction, appointed to examine arithmetic texts for use in the schools. He praised the committee for their call for simplicity and clarity. A summary of this talk was published in May [204].

On 2 March the Academy of Sciences discussed proposed reforms in the Gregorian calendar. The government had requested their opinion, and Somigliana and Silva had been appointed to study it. They reported that they were opposed to too radical reforms. Peano also made a few remarks on this occasion, and their report was then approved by the Academy. Peano would later become more closely involved in this question. On 16 March, Peano gave to the Academy a copy of his newly published book on arithmetical games [203]. At the 27 April session, it was agreed to let Vito Volterra represent the Academy at the Toronto congress; it was known that he would go (and besides, he was a senator.) He had been invited to give one of the principal lectures, but in the end he did not go, and Peano was the official delegate of the Academy of Sciences.

Before that congress, however, Peano attended another, the International Congress of Philosophy, in Naples, 5–9 May 1924. One of the major topics at this congress was the question of relativity and two days were devoted to it. Peano's knowledge of this subject was very limited, but he joined the

discussion to announce the publication that year of the book by his friends
Boggio and Burali-Forti criticizing the theory of relativity. It was an unfortunate intervention. (They claimed, in their book, to have proved the
impossibility of relativity. Peano, too, to the end of his life remained opposed
to the theory of relativity.)

Before leaving for the congress in Toronto, Peano prepared a propaganda
pamphlet for Interlingua [205] − and no doubt took copies along with him.
He went to New York by ship and (presumably) by train from there to
Toronto, taking a young relative, his *pronipote* Mario, with him. On other
trips, Peano's wife often accompanied him; he sometimes took other relatives.

There were eleven Italians in Toronto for the International Congress of
Mathematicians, 11−16 August, 1924. Peano and Guido Fubini were the only
two from Turin, and Peano was the official delegate of the Academy of
Sciences, the Academia pro Interlingua, and the University of Turin. After
the morning session of 11 August a group photograph was taken. Peano is
near the center, in the last row, with vest buttoned, but coat unbuttoned, his
tie skewed to his left. He is balding, but has a full grey beard.

Only the abstract of Peano's paper 'On equality' was published − but it
is probably almost all he said [206]. He later reported his experience in an
Academia pro Interlingua circular [206′]:

On 12 August I read my study 'On equality' in the Interlingua *Latino sine flexione*,
before members of the congress with various languages; the majority were Americans,
speaking English. The whole public, which in general was unaware of the existence of
any interlingua, declared that they understood me. The result was marvelous; but the
majority of those present were of the opinion that the English language was the language
of the future.

The tone of disappointment is obvious, but there were no doubt some who
appreciated Peano's efforts. One of these was probably Count Alfred Korzybski, who presented a paper on 'Time-binding: the general theory.' He gave
Peano a copy of his book *Manhood of Humanity. The Science and Art of
Human Engineering*, inscribing it: "To Professor G. Peano with deep respect
and admiration."

The lectures and papers were interrupted on the fourth day (14 August)
by an all-day outing to Niagara, which included an inspection of the generating
station at Queenston on invitation of the Hydro-Electric Power Commission
of Ontario. Luncheon was at the Clifton Inn, Niagara Falls, as guests of the
Power Commission, and the party returned to Toronto by boat.

Peano probably celebrated his 66th birthday aboard ship on the return home from New York. He was now among the older professors at the university, the yearbook for 1924–25 listing him as seventh in seniority among the *stabile* professors. This category seems to have (temporarily) replaced *ordinario*. This yearbook also contains an oath of allegiance to be taken by professors upon their nomination:

I swear to be faithful to the King and to his Royal successors, to loyally observe the Constitution and the other laws of the State, to exercise my office as teacher and carry out all my academic duties with the purpose of forming citizens who are industrious, upright, and devoted to the Fatherland.

On 13 November he wrote a letter to the editor of the *Bollettino di Matematica* (which was published the following year) [207.1] in which he adds several items to the bibliography of an article on numerical approximations by L. Siriati. They were all written by himself, Elisa Viglezio, and Ugo Cassina. Peano noted that Cassina was then "assistant for algebra" at the University of Turin, but by the time of publication, the editor added that he was now a professor at the Naval Academy in Livorno.

Peano attended none of the sessions of the Academy of Sciences during January, February, and March of 1925, but on 25 January it was announced that some material had arrived which referred to the proposed calendar reforms. It was to be given to Peano who was on the Committee for Reform of the Calendar "as representative of our Academy." It appears, then, that Peano had been appointed to the national committee.

The Mathesis Society held its congress this year in Milan, 29–31 October. Peano presided over the afternoon session of Thursday, 29 October. On Sunday, 1 November, there was an outing to Bellagio and Como; Peano appears in the group photograph taken on that occasion.

As was mentioned earlier, the arrival of F. G. Tricomi in the fall of 1925 (his change of address being duly noted in the membership list of the Academia pro Interlingua, of which he was a member for several years — something he later seems to have conveniently forgotten!) profoundly affected Peano's schedule, for, while nominally remaining Professor of Calculus, he in fact taught the recently instituted Complementary Mathematics. His assistant, also nominally 'for calculus', was now Clementina Ferrero, a 1924 graduate in mathematics 'with honors'. (The yearbook lists her seven publications and, curiously, six of them deal with agriculture!)

This year the rector of the university proudly reported that the Rockefeller Foundation had made several grants to the University Institutes of General Pathology, Physiology, and Anatomy. Peano's contacts with America were also increasing: in an Academia pro Interlingua circular dated 1 August, 1925 [207.2] he announced the formation of the International Auxiliary Language Association.

The IALA (as it was usually called) was founded in 1924 under the leadership of Mr. and Mrs. Dave Hennen Morris of New York (who also furnished the necessary funds.) Its purpose was "to promote widespread study, discussion, and publicity of all questions involved in the establishment of an auxiliary language, together with research and experiment that may hasten such establishment in an intelligent manner and on stable foundations." Its first president was Dean Earle B. Babcock (New York University), who held that post until 1936. Mrs. Morris became Honorary Secretary of the Association and later also Chairman of the Research Division. Mr. Morris was IALA's Treasurer from the founding of the Association until his death in 1944. (He was ambassador to Belgium, 1933–37.) Mrs. Morris also became a member of the Academia pro Interlingua and by 1927 was a member of its Directive Council.

In 1925 Peano dropped the name 'Circular' from the 'regular' publication of the Academia pro Interlingua, so that the title was simply *Academia pro Interlingua* (although it was sometimes referred to as the *Rivista* of the Academia pro Interlingua.) A total of 112 pages was published in 1925, which included the first installment of Peano's history of Interlingua: 'Volapük' [208]. Two more installments appeared the following year: 'The Academy in the period 1893–1908' [209] and 'Simplified Latin' [210]. These were then followed by a brief note on 'Chinese' [211].

The academic year 1925–26 was the first year since Peano's election to the Academy of Sciences in 1891 that he did not present a note, either by himself or by another. He was, however, called upon twice this year to pass judgment on others' work. On 29 November, 1925, he was asked to examine and report on a request of a Prof. Schow (of Trieste) for a judgment of his note "treating an argument of higher mathematics." On 13 December, Peano reported that he had already written Prof. Schow his opinion, and had been thanked by him. Again, on 21 March, 1926, the president proposed and the Academy approved, asking Peano to examine and report on a note by the Polish mathematician T. Ciuropajlowicz dealing with Fermat's Theorem. (Peano was not present at that session.) Peano reported on 11 April that he "did not think it opportune to publish the note." Having once published that Lindemann had proved Fermat's Theorem, presumably he was not to be fooled again!

In the university yearbook for 1926–27, the rector in his report mentioned

Mussolini (without naming him) for the first time, finding cause for great jubilation in the "miraculous protection of Providence" that four times the previous year saved him from attempts on his life.

The military was also beginning to have more influence on the university as a course in military history was begun the previous year for all students. Already the year before that, a series of lectures had been given by professors from the Military Academy, and the University of Turin was proud to have been the first to hold such a series of lectures.

The first volume of *Schola et Vita* (*Organo de Academia pro Interlingua*) appeared in 1926, with Nicola Mastropaolo (in Milan) as director. Peano contributed two notes to this first volume: 'Magic squares' [212] and 'Arithmetical games' [213]. Peano's devoted disciple Ugo Cassina also played a key role in the development of *Schola et Vita* and became its editor.

At the Academy of Sciences on 12 December, there being no papers presented or other business, Peano took the occasion to plead for an increase in the number of pages that could be given to non-members. The President noted that this would have to be resolved by the Adminstrative Council, but in the meantime reminded Peano that the Treasurer's report on the cost of publishing notes by non-members ("perhaps already too optimistic") would make it difficult to accede to Peano's request. Still undaunted, Peano then observed that the Academy ought to buy a new edition of the Encyclopedia Britannica. He pointed out that the copy the Academy had was the edition of 1888 and that the other libraries in Turin only had old editions. The 12th edition was then being printed; Peano reckoned the price to be no more than 2600 lire. Another member backed him up, adding that the Academy, as a scientific organization, should be able to get a reduction in price. This time the President promised to take the matter to the Administrative Council.

In the first issue of the *Academia pro Interlingua* for 1927, Peano reviewed [214] the book *Delphos, the future of international language*, by E. S. Pankhurst, who was an admirer of Peano's work for Interlingua.

A one-page note (in Interlingua) on the etymology of the word 'mathematics' was published in a journal for middle school pupils [215]. Peano's conclusion is that the meaning of 'mathematics' is 'science of the sciences'. This was followed later in the year, in the same journal, by a similar article giving etymologies of 'arithmetic', 'geometry', 'stereometry', 'astronomy', 'music' (for the Pythagoreans 'music' meant what we now call 'acoustics'), and 'algebra' [218].

At the session of the Academy of Sciences on 24 April, 1927, Peano was asked to give a report on calendar reform, which he did at the next session on

8 May. The report was greeted with applause and led to a discussion in which several members took part. At the end, the President proposed that it be sent to the *Classe di scienze morali* (the other section of the Academy) before being published [as 216] and this was approved by the members. In the report Peano recalled that the Academy had already expressed its opinion on the subject at the session of 2 March, 1924, and he concluded: "This writer is of the opinion that we should adhere to the same proposals that have already obtained a great majority: a perpetual calendar and a fixed date for Easter." Peano followed this with a note for the *Academia pro Interlingua* on 'The history and reform of the calendar' [217].

In the fall of 1927 Piera Chinaglia became the assistant for infinitesimal calculus.

The next year, 1928, there was one American, Emilia Bongiovanni (from Philadelphia) among the seven graduates in mathematics at the university.

In July the university published a collection of writings in honor of the fourth centenary of the birth of Emanuele Filiberto, restorer of the power of the Savoy dynasty. Peano contributed a historical note on Francesco Peverone and other Piedmont mathematicians of the time [219].

In August a brief (etymological) note on 'The history of numbers' was published in three journals [220].

CHAPTER 23

THE FINAL YEARS

The first International Congress of Mathematicians following the World War was held in Strasbourg, but neither it nor the 1924 congress in Toronto were truly international. From them had been excluded members from Germany, Bulgaria, Austria, and Hungary. But at the Toronto congress the delegates from the U.S.A. proposed that this restriction be removed, and this proposal was supported by Italy, Denmark, Holland, Sweden, Norway, and Great Britain. The difficulties were not entirely cleared up, but under the leadership of Salvatore Pincherle most of the obstacles were overcome at the 1928 congress in Bologna where, after the Italians, the Germans were the most numerous. .

Peano had been listed as one of the chairmen for Section 6: Elementary mathematics, didactic questions, mathematical logic, along with Ugo Amaldi and A. Perna, but at the first session on 4 September, Amaldi announced that Peano was unable to attend "for family reasons" and proposed that the section send him a greeting. This was agreed to. On 6 September, Samuel Dickstein, who was presiding over that section, proposed sending Peano a telegram of condolence "on the death of his brother" and this was also approved. Dickstein sent the following: "Sectione VI de Congressu internationale de Mathematicos mitte ad vos expressione de proprio condolentia et augurios multo vivo." The use of Interlingua was not merely in deference to Peano; Dickstein was also a proponent of this language.

In the fall of 1928 Maria Cibrario became Peano's last assistant for infinitesimal calculus. She had graduated in mathematics in 1927 with a thesis "worthy of mention" and was later to become a professor of mathematics at the University of Pavia.

At the session of 25 November, 1928, Peano (in the name of the author) gave the Academy of Sciences a copy of Cassina's *Calcolo numerico* and praised it highly, as he also did in a note in *Schola et Vita* [221]. On 2 December he presented a note nominally written by Cesarina Boccalatte, but which Cassina considered a publication of Peano [222].

Peano was especially regular in attendance at the Academy of Sciences this year, missing only one session (10 March, 1929), and he took an active part in the discussions. For example, on 24 February the president announced the gift to the Academy of a copy of Georg Mohr's *Euclides danicus*, that was originally published in Amsterdam in 1672 (and republished in 1928 with a preface by J. Hjelmslev.) Both Peano and D'Ovidio found the book worthy of discussion. At the 23 June session he was asked to "collect the sense of the members" on the possibility of establishing "centers of intellectual cooperation."

In the spring of 1929 Peano published in Interlingua an article on international words [223] and an article commemorating the 50th anniversary of Schleyer's publication of Volapük in 1879 [224]. A summary of this last article was published as a letter to the editor of *Time and Tide* (London) in the issue of 5 July, 1929, and was later reprinted in *Schola et Vita* [224.1]. A note on Italian coins in 1929 appeared later in the year [225].

Peano's intense interest in the international auxiliary language movement shows also in his frequent correspondence this year with Alice Morris in New York and Dénes Szilágyi in Budapest. (Szilágyi became a member of the Academia pro Interlingua this year, as did Otto Jespersen and D. E. Smith.)

On 30 July, 1929, the newspaper *La Stampa* (Turin) announced Peano's election as a 'national member' of the *Accademia dei Lincei*.

In September, G. Canesi, Treasurer of the Academia pro Interlingua, made the ingenious perpetual calendar that Peano had invented. (This sold for three lire, but was sometimes given to new subscribers to *Schola et Vita*.) It consists of two concentric discs, held together in a way to allow the smaller to rotate. Each is divided into seven sectors. On the outer disc each sector has the name of a day of the week and the last two digits of year numbers. The inner disc has the names of months and day numbers. To operate it, one first aligns the year and month of interest, and then reads off the day of the week corresponding to the date sought. As constructed, it can only be used for the years of the 20th century, but a table of numbers to be added for each century allows its use for the years 0 through 2599. By this time, too, at least 38 different propaganda postcards for Interlingua had been printed, of which No. 20 gave a complete grammar of the language.

From 18 to 24 September, along with several other members of the Academia pro Interlingua (Enriques, Vacca, Cassina, Gabba, Padoa, Finzi), Peano attended the 13th congress of the Italian Society for the Progress of Science.

At the Academy of Sciences Peano attended every session of the year 1929–30. On 1 December he gave the Academy (in the name of Luigi Berzolari, its principal collaborator) a copy of Volume 1 of the *Enciclopedia di Matematiche Elementari* and praised it highly, especially for the first 80 pages on 'Logic', written by A. Padoa. (Padoa had presented a copy at the Mathesis Congress in Florence, 21–23 September 1929, saying that in the article on logic he had given "the results attained in this field by Prof. Peano and his school." The audience took the occasion to express their appreciation of Peano.) On 2 February, 1930, Peano presented a note, nominally by Fausta Audisio, but which Cassina again considered to be a publication of Peano [228]. At the same session the president read an invitation from the mathematics faculty of the University of Kazan to participate in the scientific jubilee of Prof. N. N. Parfent'ev and asked Peano to write a letter saying the Academy would. Peano knew Parfent'ev personally and was happy to do so. (This was Parfent'ev's 25th year as a member of the mathematics faculty of the University of Kazan.)

Questions of Interlingua were studied in two publications in the spring of 1930 [226, 227]. Later Peano published (in Interlingua) an elaboration of his 'Algebra of grammar' [229]. This attempt to treat philology mathematically does not go beyond a simple 'arithmetic', but is nonetheless interesting as an early attempt to apply mathematics to this field.

Silvio Pivano became rector of the university on 1 November, 1928, and one of his concerns was to regularize Tricomi's position. Peano was finally persuaded to exchange the chair of infinitesimal calculus for that of complementary mathematics, not only in fact, but officially as well. He was so listed for the first time in the 1931–32 yearbook of the university. Peano's ability, however, continued to be recognized by the Minister for National Education who, in a bulletin of 14 August, 1930, appointed Peano one of three judges for applications for the *libera docenza* during 1930–31. (The other two were Salvatore Pincherle and Mauro Picone. Umberto Cisotti and Luigi Fantappiè were named as alternates.)

In 1931, Peano published a brief note on arithmetical games [230] and his last publication, a brief exhortation to the members of the Academia pro Interlingua to continue in "freedom and unity" [231]. The last session of the Academy of Sciences he attended was that of 17 April, 1932.

The next day he sent the following postcard to Cassina in Milan:

Dear colleague,

The last issue of *Schola et Vita*, of which you are the secretary, is very well done! Good! The *Revue générale des sciences*, Gaston Dorit publisher, Paris 6, in the issue of 31 March 1932, page 185, has published a favorable review of the book of Cassinis by R. de Montessus de Ballore, *dr. es sciences*. It could perhaps be of interest to you for your *calcolo numerico*.

Greetings,
G. Peano

On Tuesday 19 April, Peano taught his classes as usual. On the way home he stopped to see the film *Il Tenante Allegro* (*The Smiling Lieutenant*, with Maurice Chevalier, Claudette Colbert, and Charles Ruggles) and at dinner that evening he told how much he enjoyed it. He had recently been bothered by a cough and his wife prepared a camomile tea to help mitigate it. This night his cough was worse than usual and she wanted to call a doctor, but he refused. About 2:30 a.m. a coughing fit awakened him. Nothing seemed to help and a doctor was immediately sent for, but Peano's condition continued to worsen and the doctor arrived too late to help. Peano died around 3:30 the morning of Wednesday 20 April, 1932. *La Stampa* reported the next day that he was a victim of "an attack of *angina pectoris*, probably complicated by internal bleeding."

CHAPTER 24

AFTERWARDS

Despite the opposition at the university from F. G. Tricomi and others, Peano's last years were serene and satisfying. He was a respected member of the Academy of Sciences, for example, and of course a revered figure in the international auxiliary language movement. After his death, testimonials arrived from all parts of the world that equalled those nearer to home.

Guglielmo Marconi, as President of the *Accademia d'Italia* and the *Consiglio Nazionale delle Ricerche* sent condolences, as did Tullio Levi-Civita of the University of Rome. Samuel Dickstein of the University of Warsaw wrote: "I have received with deep grief the news of the death of our illustrious president, Prof. Peano, a sublime and generous man, a dear and venerated friend." Sylvia Pankhurst notified Bertrand Russell, who wrote her: "He was a man I greatly admired from the moment I met him for the first time in 1900 at a Congress of Philosophy, which he dominated by the exactness of his mind." Mrs. Dave H. Morris, honorary secretary of the International Auxiliary Language Association, wrote from New York: "His altruistic work and his generous spirit for the cause of an international language will remain as example and inspiration for the workers in this cause."

The funeral was held on 22 April from 10:00 to 12:00. The body was accompanied by Count Thaon di Revel, mayor of Turin; Prof. Pivano, rector of the university; General Pintor, commandant of the Military Academy; Prof. Somigliana, president of the faculty of sciences of the university; Prof. Parona, president of the Academy of Sciences; and Prof. Mattirolo, director of the Botanical Institute. The procession moved from his house, in Via Barbaroux 4, to the cathedral where a funeral mass was celebrated, then across the Piazza Castello and along Via Verdi to the entrance hall of the university where, according to the academic custom, the final farewell was said. Pivano, Somigliana, and G. Canesi (director-treasurer of the Academia pro Interlingua) spoke. Afterwards, the body was taken to the city cemetery, the *Cimitero Generale*, where, at Peano's expressed wish, he was given a simple burial, in the common field No. 3 south, in grave No. 674.

In his funeral oration, Pivano emphasized Peano's "enthusiasm for learning and his love for young people; the simplicity and integrity of his life." Somigliana saw him as "a man of science and nothing except a man of science." Canesi looked to the future and predicted: "The victory of your Interlingua will make your name immortal."

Obituaries of Peano were published in a number of newspapers and journals in Italy and abroad. In New York, the International Auxiliary Language Association commemorated Peano at its session of 28 November, 1932. It was pointed out that Peano was in contact with IALA from its beginning and that even though he was unable to attend their international congress in March—April 1930, he was in frequent contact through almost daily communication. (Peano was denied a travel permit by the fascist government.)

In Odessa the Scientific Research Department of Mathematics held a special session in his memory on 16 December, 1932, at which D. A. Kryzhanovskii, who had studied in Rome, spoke about Peano's mathematics.

Peano's death was announced at the Academy of Sciences on 24 April by Parona, who proposed that a commemoration of Peano be entrusted to Boggio "who was a beloved disciple of Peano and a devoted and affectionate friend." D'Ovidio then asked for the floor and gave an emotional little speech remembering Peano. All of this was heard standing by the members, who then gave to Boggio the task of writing a commemoration. This he did, but it was not read until a year later, on 18 June, 1933. At the *Accademia dei Lincei* in Rome, the president announced Peano's death and briefly recalled his merits. He also expressed the desire that a commemoration be held "in one of the next sessions," but the *Acts* does not record that this was done. (*Schola et Vita*, however, reported that Peano was commemorated there by F. Severi.)

Indeed, after a flurry of writings recalling Peano's merits, general interest in his mathematics nearly died out. A few months after his death, the faculty of sciences at the university considered the possibility of publishing a selection of his writings and appointed a commission consisting of Carlo Somigliana, Guido Fubini, and F. G. Tricomi, who worked out a project in 1933. The presence of Tricomi on this commission practically guaranteed, however, that nothing would come of the project, and in fact the project was abandoned until after the Second World War when, Tricomi being in the U.S.A., an analogous project was again planned by T. Boggio, G. Ascoli, and A. Terracini. In the meantime the *Unione Matematica Italiana* had decided to publish Peano's works, but delayed so as not to interfere with the plans of the university. The latter, however, abandoned this project in 1956 (Tricomi had in the meantime returned to Turin) and so the UMI then asked Ugo Cassina to propose a project for publishing Peano's works and on 5 October, 1956, named a commission consisting of Giovanni Sansone, president of the UMI, A. Terracini, and U. Cassina to make the final selection of works to be published.

Three volumes of 'Selected Works' were planned and these appeared, under the editorship of U. Cassina, in 1957, 1958, and 1959. To these was added in 1960, with the financial help of the city of Cuneo, a facsimile reprint of the last edition of the *Formulario Mathematico*, with an introduction and notes by Cassina. With the publication of these volumes, making the scattered and out of print works of Peano again readily available, there has been a renewed interest in his accomplishments, especially among historians of mathematics.

Public recognition has also come to some extent. In Cuneo, the *Liceo Scientifico* was named after him in 1953 and a bronze bust of him by G. B. Alloati was placed in the entrance hall. In 1963, Peano's remains were to be removed from the grave in Turin and the question arose as to what to do with them. The Peano family decided to have them transferred to the family tomb in Spinetta (Cuneo) and the mayor of Cuneo insisted on a small ceremony for the occasion. The city donated a marble plaque dated 31 October, 1963.

Peano's high hopes for the success of the international auxiliary language movement seem farther from realization today than they were at the time of his death. Even the name 'Interlingua' has been appropriated by others. This last would not have bothered him — he once said to Sebastiano Timpanaro: "What does it matter that an idea succeed with the name of Peano or of another? The important thing is that it succeed." But he would surely be disturbed by the lack of success of this language.

On the other hand, some of Peano's ideas in logic and mathematics had by mid-twentieth century borne real fruit, and in the second half of this century their importance has been seen more and more clearly. Peano would probably be astonished at the role that has come to be played by infinite-dimensional vector spaces, for which he gave the first axiomatic treatment. But in doing that he was ahead of his time, and his work was ignored for many years. He would be surprised that his name has become attached to several things, for example 'Peano spaces', that he knew nothing about.

Peano was, as Somigliana said in his funeral oration, "a man of science" and he will be remembered for this. He will also be remembered for being a man of character — honest, modest, and never vindictive. Among interlinguists — notorious for their rivalries — "it is remarkable," as J. Chanaud, proponent of another language wrote, "that in the whole work of Prof. Peano one does not find one single word of attack, aggression, or sharp polemic against any individual or system. Is that not a beautiful and memorable example?"

CHAPTER 25

SUMMING UP

The preceding strictly chronological account of the life and works of Giuseppe Peano is based on my efforts over a decade to find and record the facts of Peano's life. In writing it I have often felt myself to be more a writer of chronicles than a historian. But I have been especially concerned to 'let the facts speak for themselves,' since the facts have simply not been readily available earlier, although Peano has been dead for nearly fifty years. The reason for this lack was not obvious to me at first and only gradually became clearer. In this concluding chapter I shall try to point out some of the major themes in his work and suggest some of the reasons for the neglect of his work by historians of mathematics.

Peano lived too long. Had he not lived beyond the 19th century, I am sure I would not be writing this account, for his work would already have been discussed in detail and his fame would be far greater than it is. Like many mathematicians, though by no means all, Peano did his best work in his early years. Indeed all his major accomplishments were in the 19th century; he lived well into the 20th century, but without the brilliant mathematical career his early work anticipated.

To say that Peano's later work does not equal the earlier does not, however, tell the whole story; outside forces operated to obscure his reputation. After Peano's death, while the rest of Europe waited for Italy to evaluate her own, the rest of Italy expected this from Turin. In Turin, however, the anti-Peano forces were in firm control. As late as 1967, during a conference I attended there commemorating Federigo Enriques, a professor leaned over to me and whispered, "Listen to the anti-Peano sentiment!" Even later, while President of the Academy of Sciences of Turin, F. G. Tricomi continued to publicly make anti-Peano statements. I have earlier discussed some of the causes of the antagonism to Peano on the part of the mathematics faculty in Turin, but the continued attacks on his reputation thirty-five years later are inexplicable.

Of course, Peano had his defenders. The most knowledgeable and the most devoted of them was Ugo Cassina, whose admiration and affection for Peano lasted a lifetime. He was, however, never in a position of power and his strong, at times overzealous, defense of Peano was not taken seriously, partly because he also shared Peano's zeal in promoting Interlingua. We are indebted to Cassina for many studies of Peano's work as well as the three volumes of the *Opere Scelte* and the reprint edition of the *Formulario Mathematico*. But Cassina's articles were often meant to be more than strictly historical studies. Cassina apparently hoped to keep alive Peano's mathematics and logic, and in this he was not successful. Nor was he concerned with sharing his more personal knowledge of Peano; he refused, for example, to answer my direct question as to whether Peano regretted not having children of his own.

It has, in fact, been difficult for me to form more than a general impression of Peano's personality. I have been in contact with a number of persons who knew him personally, but that was many years after his death. The picture that emerges is of an uncomplicated, non-aggressive, egalitarian figure. At a time when professors traditionally held themselves aloof from their students, Peano did not hesitate to invite them to have an ice cream with him, and he always paid. Politically, he was a socialist, but in a non-partisan, rather romantic 19th century fashion. "Woe to him who only reads one newspaper!" he would often say.

Peano's scientific career may be divided into two periods and, conveniently, the dividing point is the end of the 19th century. The earlier period contains most of his significant achievements in the three major areas of analysis, foundations, and logic. In the list of his publications which Peano had printed in 1915 (probably meant as a defense of his teaching) he wrote that his major work had been in "calculus". Let us review some of those accomplishments first.

Peano was a master of the counter-example. The most spectacular of these was his space-filling curve, which showed that a continuous curve cannot always be enclosed in an arbitrarily small region. His ability to find an analytic expression for such a function was earlier shown in the case of the function defined to be 0 for rational numbers and 1 for irrational numbers. The space-filling curve advanced mathematical theory directly, but Peano's many counter-examples contributed greatly to advances in the teaching of calculus. The space-filling curve is well known by Peano's name; one forgets, for example, that the function which perhaps most calculus texts give as an example of a function whose second partial derivatives do not commute appeared first in a text by Peano. His two calculus texts, especially the 'Genocchi', had a wide influence on the trend toward more rigorous teaching of the subject.

Peano's theorem for differential equations, although not always named after him, must be regarded as fundamental. Today the theorem, that $y' = f(x, y)$ has a solution on the sole condition that f be continuous, is usually seen as a special case of a more general fixed point theorem, but Peano's 'elementary' proof continues to arouse the interest of teachers. Peano's independent discovery of the method of successive approximations for the solution of differential equations must be considered as remarkable, although he was not the first to use this method (as he and Cassina believed).

Even Bourbaki now refers to 'Jourdan—Peano' measure, although Peano's independent discovery of it was long ignored. Lebesque, however, acknowledged Peano's influence on his own developments. Peano's use of upper and lower bounds in defining areas and integrals did not become part of standard textbook presentations, but may be seen as part of a general trend among mathematicians toward a wider use of topological considerations. On the other hand, Peano's anticipation of later developments in his study of functions of sets was not continued and had little direct influence.

Perhaps the most widely known of Peano's accomplishments is his set of postulates for the natural numbers. Very early, however, his originality was denied. I have elsewhere tried to explain this, but it may be repeated here that we have Peano's own word that he discovered the postulates independently of the work of Dedekind. I see no reason to doubt him. The question was 'in the air' of course, but that should not take anything away from our estimate of his originality or his ability to cut through traditional complications to arrive at his simple statement.

Peano played a leading role in the late 19th century movement toward the axiomatization of mathematics. Besides the axioms for the natural numbers, there was a parallel, if not as original, axiom system for geometry. In this context, too, may be recalled Peano's use of recursive definitions in his development of arithmetic from the axioms. Somewhat slighted at the time, recursive definitions have had their importance widely recognized in the 20th century.

With his axioms for a linear space Peano was far in advance of current developments. Here, again, he had no direct effect on the later extensive development of this subject. He was, however, influential in promoting the use of vectorial methods in Italy, if not always able to overcome opposition to their use.

There is no question of Peano's pioneering role in the development of mathematical logic. This was widely recognized at the time and the debt was duly acknowledged, especially by Bertrand Russell. In the second half of the 20th century there has been a re-evaluation of the work of Frege, which shows him to have been a subtler thinker than Peano in this field. Frege's influence on his contemporaries, however, was minimal; here Peano was definitely in the mainstream. If Peano was unsuccessful in distinguishing theory from metatheory, he was able to point out distinctions, such as that between a set containing one element and that element, which eluded others. Indeed, much of his symbolic writing was survived and is in general use today.

The accomplishments reviewed above all took place in the 19th century. Because of this, the fact that the level of accomplishment did not continue into 20th century is inevitably disappointing and the question arises: Why was this so?

Apart from general considerations of age and personality several answers may be given to this question. One is that Peano was a mathematical opportunist. He sought out fields in which one could get immediate results. When these were obtained, he moved on to more fertile ground. By 1900, having spread his interests over several fields of mathematics and logic, he was unable to give any one the persistent attention demanded to produce continuing results.

A second reason is that his editorial activities, especially with the *Formulario*, not only distracted him from original research, but also drew his attention away from new mathematics to a cataloguing of old results. This situation was probably intensified as work on the project became more and more a one-man operation.

Perhaps a third reason for Peano's later lack of original work was his alienation from more productive members of the faculty of mathematics at the University of Turin. Whatever the cause, this took away the normal stimulation to mathematical activity that arises from such contact.

A fourth and perhaps most important reason for Peano's later lack of scientific activity in general was his dedication to the international auxiliary language movement. To be sure, Peano approached the task of constructing his Interlingua more scientifically than others had done, but the result was no more viable. Peano's idealism and zeal in this cause must be admired, but his goal seems no closer now than when he died, and I am very skeptical that it can ever be realized.

Personally, I find the second period of Peano's career the more interesting period in human terms. This is the man I would like to have known. I am fascinated by his gentle personality, his ability to attract lifelong disciples, his tolerance of human weakness, his perennial optimism. At the same time, as a historian of mathematics, I recognize the 19th century period as the more important period of his scientific career for the development of mathematics and logic. Thus I believe Peano may not only be classified as a 19th century mathematician and logician, but because of his originality and influence, must be judged one of the great scientists of that century.

PEANO'S PROFESSORS

Enrico D'Ovidio was only 33 years old when Peano became a university student, and he was to survive Peano. He was born on 11 August, 1842, in Campobasso, then in the Kingdom of the Two Sicilies. D'Ovidio studied in Naples with Achille Sannia in preparation for entrance into the School of Bridges and Roads, where he also studied for some time. He was influenced to take up an academic career, however, by the lectures of Battaglini, Fergola, Del Grosso, and Padula, and indeed the first volumes of Battaglini's *Giornale di Matematiche* (founded in 1863) contained a few notes by D'Ovidio on determinants and conics. He did not pursue a regular university course, but after teaching in secondary schools, he was granted a degree in mathematics *ad honorem* by the Facoltà di Scienze di Napoli on his application. He was dispensed from any examination, it being considered that his scholarship was already sufficiently proven. In 1869 he published, with Sannia, a geometry text for use in the classical schools. The teaching of Euclid's *Elements* had recently been prescribed for these schools and this text remained important for some time.

Although D'Ovidio did not wish to leave his family and native region, he was encouraged by Eugenio Beltrami to enter the competition in 1872 for the chair of Algebra and Analytic Geometry at the University of Turin, where he spent the remaining 46 years of his life. His most important scientific work occurred during his first 20 years there, but even after his retirement in 1918 he continued to take an active role in the life of the university and the Academy of Sciences. His presence in Turin, especially in the early days, contributed greatly to the improvement of North-South relations, both at the university and in the artistic circle presided over by Edmondo De Amicis and Giuseppe Giacosa.

In the few years preceding 1872 the work of Lobachevskii, Bolyai, and Riemann in non-Euclidean geometry was becoming well known, as well as the *Sixth Memoir* of Cayley on projective metrics and Plücker's idea of considering lines as fundamental ideas in space. An important group of papers by D'Ovidio dealt with these ideas, especially with Euclidean and non-Euclidean projective metrics, culminating in the memoir of 1877 on "The fundamental metric functions in spaces of arbitrarily many dimensions with

constant curvature." At the same time, D'Ovidio was investigating binary
forms, which influenced Peano's early publications. Already in 1876, D'Ovidio
had published the first part (dealing with conics) of a three-part text in
analytic geometry (the other parts were: quadrics in 1883 and the analytic
theory of the fundamental geometric forms in 1885) which displayed his
didactic ability. Because of D'Ovidio and his student Corrado Segre, an
important school of geometry was developed at the University of Turin.

D'Ovidio was instrumental in the establishment of the post of assistant
for the exercises in algebra and analytic geometry. He always tried to have his
best graduates for this position, so that it came to be a coveted position. His
first assistant was Fortunato Maglioli (1878–79). Then came Francesco
Gerbaldi (1879–80). Peano was D'Ovidio's assistant from 1880 to 1883. He
was followed by Corrado Segre (1883–84), when Peano became assistant
to Angelo Genocchi.

D'Ovidio was chairman of the Facoltà di Scienze in 1879–80 and rector
of the university from 1880 to 1885. He was again Science Chairman from
1893 to 1907, leaving this post (as well as that of Director of the Mathematics
Library) when he was named a commissioner of the Polytechnique of Turin,
of which he was later director until 1922. He was a member of the principal
Italian scientific societies (the Academy of Sciences of Turin from 1878, the
Accademia dei Lincei from 1883, the Society of the Forty from 1884.)
He was named a senator in March 1905, but there were rumors that this
was due to a mix-up and that the nomination was intended for his brother
Francesco, the noted philologist, who was in fact also named senator a few
months later. D'Ovidio died in Turin on 21 March, 1933.

Peano's professor of design, *Count Carlo Giulio Ceppi,* furnished a contrast
to D'Ovidio. He was older (he was then 47), affected mannerisms, and was
not at all a scientist — more a schoolteacher than a professor. He was born in
Turin on 12 October, 1829, and died there on 9 November, 1921. He studied
architecture with Carlo Promis at the University of Turin and received a
degree in civil engineering and architecture on 12 August, 1851. Since he did
not lack money, he decided to work on his own, but after the death of his
mother his father urged him to take a regular position. To please him, Ceppi
obtained in 1857 the vacant post of design teacher at the Royal Military
Academy. There were probably other reasons for taking this post: the work
required little preparation, there were many holidays, and he was dispensed
from service in the National Guard. He became prominent in 1863 when he
won first prize for his project for a facade for the Church of Santa Maria del

Fiore in Florence. (This project, however, was never carried out.) He taught at the Military Academy until 1870, but already in 1869 he succeeded Promis as teacher of architecture at the School of Engineering, which later became the Polytechnique. Shortly afterwards, for reasons not entirely clear, he transferred to the university. Beginning around 1867 a number of buildings were erected after his design, perhaps the best of these being the Palazzo Ceriana, designed in 1878. Ceppi was noted for never wearing a topcoat and it is said that from the age of 20 he slept, summer and winter, without covers except for a single sheet.

Donato Levi was five years younger than Ceppi and, like Ceppi, a graduate in civil engineering and architecture (1856). From the age of 18 he was an assistant at the Astronomical Observatory of Turin, where he was principally concerned with meteorological observations. Before this he had already started his teaching career. After five years as a secondary school teacher of mathematics, he taught privately and in 1866 he won the competitive post of *dottore aggregato* in science at the University of Turin (a rank carrying no salary or duties.) From 1875 he was assistant in the school of projective and descriptive geometry. His published works are few and not very important, but they display that orderliness, clarity, and precision of language which was valued in his classroom lessons. Levi was precise and punctual in all his duties until December 1883, when the sickness began that, after several improvements and relapses, carried him off on 21 August, 1885, at the age of 51.

Hugo Schiff, like Levi, was 43 years old when Peano attended his first class in chemistry. This was also Schiff's first class in Turin. He was born on 3 April, 1834, in Frankfurt am Main, into an old Jewish family that traced its origin to Spain. He received his chemical training in Göttingen, but transferred to Berne in 1856 for political reasons. He returned to Göttingen to obtain a Ph.D. in 1857, but returned again to Berne, where he began teaching chemistry in 1858. In 1862 he and his brother Moritz went to Italy where he worked first in a laboratory in Pisa, then was professor of chemistry at the Museum of Natural History in Florence, then professor at the University of Turin (1876–79), and finally professor at the University of Florence from 1880.

Schiff liked to tell about the sacrifices he made during the first years of his career, especially in Switzerland, but he was very economical and after 50 years he was able to leave a large inheritance. Politically, he was a socialist and one of the founders of the newspaper *Avanti*. An accomplished linguist, he knew several modern European languages, as well as Greek, Latin, Arabic, and Hebrew. He was a member of the Academy of Sciences of Turin from 1898.

Schiff had notable success in his study of organic chemistry. In his laboratory teaching he was very serious, taking special care in his teaching of chemical analysis. He also invented a number of useful laboratory instruments and published an excellent introductory text in chemistry in 1876. Schiff enjoyed good health and worked hard until December 1899, when he fell in the street and was run over by a loaded cart. He never recovered completely and died in Florence on 8 September, 1915.

Michele Lessona was born on the outskirts of Turin on 20 September, 1823, the son of Carlo Lessona, a professor of veterinary science. He enrolled in the medical program of the University of Turin, obtained a post in the Collegio delle Provincie, and was graduated on 12 August, 1846, with a degree in medicine and surgery. For a brief period he practiced medicine in Turin, but he soon fell in love with one of his sisters' teachers, with whom he fled to Egypt to escape the anger of her parents, who were opposed to their marriage. There followed a period of wandering, but after stops in Malta and Cairo, he was named director of the hospital in Khankah. He left Cairo, with his wife and baby only a few months old on 13 May, 1848, but after his arrival in Khankah, a cholera epidemic broke out that took his wife, leaving him alone with his daughter. When the epidemic was over he returned to Cairo and near the end of 1849 to Turin.

From 1850 to 1854 Lessona taught in secondary schools in Asti and Turin. He was then given the chair of mineralogy and zoology at the University of Genoa. During this time he published a series of notable secondary school texts in natural history, physical chemistry, and zoology. He had long wanted to make an extended trip and was finally able to satisfy this desire in 1862, when he was appointed physician to a diplomatic mission to the Persian emperor.

During the academic year 1864–65 Lessona was professor of zoology at the University of Bologna. The following year he returned to Turin as substitute for De Filippi, where after the latter's death he was named professor in April 1867. During this period he continued publication of original researches, popular scientific works, and translations, the most important of which was Darwin's *The Descent of Man*.

Lessona was several times elected to public office in Turin and was rector of the university from 1877 to 1880. In 1889 he was named president of the

Academy of Sciences of Turin, of which he had been a member since 1867. On 21 November, 1892, he became a senator. His death came on 20 July, 1894.

Giorgio Spezia was born on 8 June, 1842, in Piedimulera (Province of Novara) and was buried there after his death on 10 November, 1911. He began higher studies at the University of Pavia, but transferred to Turin where he received an engineering degree in 1867. He continued mineralogical studies in Turin and, from 1871 to 1873, at the universities of Göttingen and Berlin. From 1874 he taught mineralogy at the University of Turin, being named professor in 1878. His researches were rewarded in 1893 and 1900 with prizes from the *Accademia dei Lincei*.

Spezia was good-hearted, although with rather plain manners. He avoided society, preferring the quiet of his laboratory. He was a member of a number of academies, including: the Academy of Sciences of Turin, the *Lincei*, the XL, the Imperial Mineralogical Society of Moscow, and the Italian Geological Society.

Angelo Genocchi was born in Piacenza on 5 March, 1817. He had all his schooling there, graduating from the university in 1838 with a degree in law. After several years of practicing law, he became professor of Roman Law at the university in 1845. This was a position he had not sought, but he accepted it when offered – to the regret of some of his early students, who found him too cold and severe. There was warmth, however, in his political feelings and already before 1848 he was part of a group formed to combat Austrian clericalism. With the uprising in Lombardy in March 1848 the Austrians vacated Piacenza and the liberals formed a provisional government whose principal aim was the annexation of the city and province to Piedmont. But events moved too quickly for this and the Austrians returned to Piacenza in August. Genocchi was greatly shaken by this turn of events. Before the Austrians could return victorious, he and several other liberals left the city. After settling in Turin he continued to receive offers to return to Piacenza, but he refused these, saying that he would not return until freedom returned. Genocchi ceased his political activity, however, with his arrival in Turin and turned his energies to mathematics.

In Piacenza he had studied many subjects and was well prepared in law, but his favorite study was always mathematics. He found in Turin many incentives toward a more intensive study of mathematics. Perhaps, too, his personality was not well suited to the practice of law. He was not a good

debater. He quickly lost his patience, would shake and become bitingly bitter
— but he quickly grew calm again. He led a simple, retiring life and mixed
little. At a period when mathematics in Italy was at a low point, Genocchi
contented himself with the frugality necessary to his small income and
devoted himself to the study of mathematics. He frequented the libraries,
attended the lectures of Plana and Chiò, and in 1851 published his first article
in mathematics. He began teaching, not before 1857, urged to it by Chiò,
who cut short Genocchi's hesitation by writing and signing Genocchi's name
to the application for admission to the competition for the chair of algebra
and complementary geometry at the University of Turin. He won the competition and took over the duties, being given regular title to it in 1859. In
1860 he transferred to the chair of higher analysis (later abolished), in 1862
to Introduction to the Calculus, where he was sure to have a good number of
students, and in 1863 to infinitesimal calculus, where he remained until
Peano substituted for him in his later years.

The qualities that stood out most in Genocchi as a teacher were learning
and precision. Although devoted to his research, he never neglected his
teaching duties. He was scrupulously punctual and justly demanding of his
students. They knew this and feared him, but he liked them and never held a
grudge, even after severely rebuking some during the exams. His explanations
were calm, with no repetitions, and he aimed at rigorously presenting the
fundamental concepts and studying them so as to arrive at simple procedures
and clear exposition. In this he succeeded quite well. He was, in fact, one
of the first to infuse into the teaching of calculus that spirit of exactitude
which is common today, and he did it without giving up clarity. He was not,
however, a perfect teacher. His austere character, along with his thin and
monotonous voice, did not warm the classroom or allow the students to feel
at ease with him.

Genocchi gave great energy to his publications. From 1851 to 1886 he
published 176 scientific works in Italian, French, and German journals. These
referred especially to number theory, series, and integral calculus. He did not
adopt the methods of Riemann and Weierstrass, but rather worked in the
tradition of Euler, Lagrange, Gauss and Cauchy. His major work was in
number theory, of which he was the principal investigator in Italy.

Genocchi kept up a frequent correspondence with many mathematicians,
both within and outside of Italy, most frequently with Bellavitis. He never
married, indeed he seemed to avoid contact with women, but he used to
advise young scientists to do so.

In his later years, Genocchi went out seldom because of his health. Among the reasons, he broke his kneecap in 1882, which left a certain stiffness in the knee. But around 1885 he began to recover somewhat and resumed his lessons (which had been taken over by Peano.) He also accepted the presidency of the Academy of Sciences. His recovered health did not last, however, and he once more gave up lecturing, gave up writing letters, and almost stopped reading them. Death came on 7 March, 1889; he was 72 years old.

The name *Eligio Martini* appears in the *Carriere dello Studente* as the person responsible for Peano's course in calculus, where one would have expected the name Angelo Genocchi. It is not clear to what extent Martini was responsible for the course. Indeed, he is an 'almost unknown.' This is the description of F. G. Tricomi, who has, however, managed to discover a few facts about him. He was born in Turin, date unknown, and died there on 1 March, 1896. In the academic year 1862–63 he took over from Plana (who was already in his 80's) the teaching of calculus, first as *incaricato* and then (from 1863 to 1865) as *straordinario*. He then moved, again as *straordinario*, to the chair of algebra and analytic geometry, which he held until the arrival of D'Ovidio in 1872. After that he is listed in the yearbooks as *dottore aggregato* (only an honorary title at that time), except that·in 1878–81 he surprisingly returns to the teaching of calculus – this time as *assistente straordinario*! He published only a few minor notes.

Giuseppe Basso was born on 9 November, 1842, in Chivasso (about fourteen miles northeast of Turin.) His father ran a tailor shop in his home and all the family worked there. It was expected that Giuseppe, too, would continue this work, but it was soon discovered that he was an excellent student and loved study, so his parents decided, at great sacrifice, to allow him to continue his education. After completing the local middle school, he continued his secondary studies in Turin. At the age of fifteen he won the competition for a post in the Collegio Carlo Alberto (just as Peano would later do) and this allowed him to enrol in the university. He received his degree in physics in July 1862, at the age of nineteen. Only a few weeks later he was named professor *aggiunto* at the Military Academy. His objective was to be self-supporting and to begin to repay his parents for all their sacrifices. To this end he accepted teaching positions in several private secondary schools. Despite his time-consuming teaching activities he was able to prepare a dissertation for the examination for *dottore aggregato* of the university, which he successfully completed in 1864. His true scientific career began with this, for this opened doors to university professors, their laboratories (which

were actually very modest), and their counsel. He soon had occasion to
substitute for Gilberto Govi in his course in experimental physics. He later
substituted for Felice Chiò, professor of mathematical physics, who died in
1871, Basso being named extraordinary professor of mathematical physics
the following year. In that year, 1872, Govi was named a member of the
International Commission for Weights and Measures and took up almost
continual residence in Paris, while keeping title to the chair of experimental
physics. But Basso was the chosen substitute. Govi gave up his position in
1876, to become director of the Vittorio Emanuele Library in Rome, but a
successor was not appointed until the end of the academic year 1877–78.
(Thus Peano's course in experimental physics was the last such course taught
by Basso. The following year it was given to Andrea Naccari.)

In 1879, the year he was relieved from teaching experimental physics at
the university, he was promoted a grade at the Military Academy and in 1881
was given title to the chair of physics there. In 1882 he was promoted to
ordinary professor of mathematical physics at the university.

All but seven of Basso's publications appeared in the *Acts* or *Memoirs* of
the Academy of Sciences of Turin, which had elected him a member in 1877
and twice elected him secretary. His publications total forty-six, the majority
dealing with optics. Eleven of them are commemorations of deceased members
of the Academy. He was elected a member of the Society of Italian Spectro-
scopists in 1891 and of the Academy of Agriculture of Turin in 1893.

Basso never married. His later life was calm and serene, his greatest satisfac-
tion being that he was able to repay his parents for their sacrifices by giving
them a comfortable old age. He was also able to support the education of
his nephews and nieces. He died suddenly on 28 July, 1895, being survived
by his mother, for whom he showed the greatest affection.

The life of *Giuseppe Bruno* resembled that of Basso in important ways.
Bruno was born of artisan parents who were unable to support his continued
education, and he, too, was able to begin university studies only after winning
a post in the Collegio Carlo Alberto. Giuseppe Bruno was born in Mondovì on
21 June, 1828. He early showed great ability and love for mathematics. With
the aid of the free post at the Collegio Carlo Alberto, he entered the university
in 1843, graduating in 1846 in positive philosophy. During the academic year
1846–47 he remained at the Collegio, after which he taught philosophy in
Ceva, returning to Turin in 1848. In the following years he devoted much
time and energy to his teaching, for he gave many private lessons in addition
to his teaching of mathematics in the Collegio Nazionale of Turin from 1850

to 1861 and in the Technical Institute from 1860 to 1891. In the meantime, he also moved slowly up the academic ladder. In 1850 he was given a degree in hydraulic engineering and in 1851 was named *dottore aggregato* in mathematics. In 1852 he was named substitute professor in mathematics, substituting until 1858 in algebra and geometry, and from 1860 to 1862 in differential and integral calculus. In 1862 he began teaching descriptive geometry and was named extraordinary professor the following year. The title was changed to Professor of Projective Geometry with Design in 1875, but he obtained permission to have his assistant teach the additional subject. (Thus Peano studied projective geometry with Donato Levi.)

Bruno stayed on this next-to-the-last rung of the academic ladder for fifteen years, finally being named ordinary professor in 1878, after the intervention of his colleagues in his behalf. They further showed their confidence in him by electing him chairman of the Facultà in 1881 and in successive years.

In contrast to Basso, Bruno was twice married. While Basso enjoyed being with his friends, Bruno had few friends and almost never left the circle of his family. Bruno was a man of duty, playing for his family, servants, and students the role of benevolent despot. But in the classroom he was an excellent teacher, both because of the clarity of his lectures and the enthusiasm with which he gave them.

Bruno's publications totalled twenty-one, all but two of them dealing with geometry, and remarkable more for the results obtained than for the method used. The small number of publications is probably due in part to his having spent so much time and energy in teaching, which despite his weakening condition he continued as long as he was able. He died of bronchial pulmonitis on 4 February, 1893.

Camillo Ferrati had held the chair of geodesy since 1852, but from 1874 he was a deputy for four legislatures and in 1876 became Secretary General to the Minister of Public Instruction. He gave up his professorship in 1880, but already had been substituted for during Peano's student years by *Giuseppe Lantelme*, who continued for another year, after which Nicodemo Jadanza was appointed. Lantelme is variously listed in the university yearbook as *ingeniere* and *dottore*, but only as substitute for Ferrati.

In contrast with the somewhat obscure figure of Lantelme is the professor of rational mechanics, *Giuseppe Bartolomeo Erba*, who was also chairman of the faculty of mathematical, physical, and natural sciences. This recognition is the more remarkable in that Erba, through an excess of modesty, never published anything!

Erba was born in Domodossola on 10 February, 1819. He completed his secondary studies in a boarding school directed by the Jesuit Order, where he was already known for his application to study and his rare modesty. It seemed for a while as if he, too, would join that order, but his father discouraged it and in 1837 he transferred to Turin and began university studies.

In 1841 he obtained a degree in hydraulic engineering and soon after a degree in architecture. He obtained the title *dottore aggregato* in 1845 and began his university teaching career on 20 October, 1850, when he was named substitute professor and charged with teaching descriptive geometry. He later taught other courses, but in 1857 he obtained the chair of rational mechanics which he held until 1891.

Erba was known for the care he took in the preparation of his course in mechanics, entirely rewriting it every year, and was generally considered a good teacher. His life was tranquil. He and his wife had no children, but he was very fatherly to his nephews and nieces. He continued to show a strong and sincere piety. Death came on 1 November, 1895.

Francesco Faà di Bruno was born in Alessandria on 29 March, 1825. At the age of 16 he entered the Royal Military Academy in Turin, receiving a commission in 1847, but he left the army in 1853. He studied under Cauchy at the Sorbonne in Paris, where Hermite was a classmate. After taking a degree he returned to Turin, where he continued his studies. In 1861 he obtained the title *dottore aggregato*. His career as a professor officially began in 1871 and in 1876 he was named Extraordinary Professor of Higher Analysis, a position he held until his death on 27 March, 1888.

After his return to Turin, Faà di Bruno devoted himself to many charitable activities (founding a home for working girls, etc.) and already fifty years old, he was persuaded by Giovanni Bosco (Saint John Bosco) to become a priest. He was ordained in Rome on 22 October, 1876. In 1881 he founded a religious order for women. Beginning about 1928 there was a movement for his canonization. This cause was officially accepted by the Sacred Congregation of Rites in 1955, and he was declared a saint on 14 June, 1971.

Faà di Bruno was tall and not always well dressed, but he was simple and good-natured. He was of a solitary disposition and spoke seldom (and not always successfully in the classroom.) He cultivated music and was said to be a good pianist. He was also the author of several useful inventions.

As a mathematician, Faà di Bruno is most known for his treatise on binary forms (1876) which had a German edition in 1881, with modifications by Max Noether.

Francesco Siacci was born in Rome on 20 April, 1839. He received a degree *ad honorem* in Rome in 1860. This was a rare honor, but his mathematical talent early showed itself and already he enjoyed the patronage of Prince Baldassare Boncompagni. For political reasons, he emigrated to Turin in 1861 and enlisted in the army there. Because of his previous studies, he was made an officer and sent to study at the Military Academy. In 1866 he took part in the war against Austria, but before that campaign was over he was recalled to Turin to teach ballistics at the Military Academy. In 1872 he was named Professor of Ballistics and he taught this subject until his retirement from active service in 1892. Already in 1871 he was called to teach mechanics at the University of Turin. He was named Extraordinary Professor of Higher Mechanics (a position created for him) in 1875 and Ordinary in 1879. He continued to teach at the university until 1893, when he transferred to the University of Naples, keeping his position there until his death on 31 May 1907.

Siacci is said to have been an excellent teacher, both at the university and at the Military Academy. He left some hundred publications, the most important being those concerned with analytical mechanics. In the application of mechanics to artillerty — ballistics — he was a master. His treatise on this subject, especially the second edition of 1888, which had a French translation in 1891, was considered masterly. He received many honors during his lifetime, including election to most of the important scientific academies of Italy. In the army reserve he reached the rank of major general.

MEMBERS OF THE SCHOOL OF PEANO

In the commemorative issue of *Schola et Vita*, following the death of Peano, Ugo Cassina published the following chronological list of Italian authors who belonged to the 'school' of Peano:

Giovanni Vailati*
Filiberto Castellano*
Cesare Burali-Forti*
Giulio Vivanti
Rodolfo Bettazzi
Francesco Giudice
Gino Fano
Mineo Chini
Alessandro Padoa
Giovanni Vacca*
Mario Pieri
Marco Nassò
Tommaso Boggio*
Angelo Ramorino*
Corrado Ciamberlini
Federico Amodeo
Filippo Sibirani
Giuliano Pagliero*
Michele Cipolla
Florentio Chionio
Margherita Peyroleri
Maria Gramegna
Matteo Bottasso

Vincenzo Mago
Eugenio Maccaferri
Alberto Tanturri
Luciano Della Casa
Angelo Pensa
Luisa Viriglio
Virginia Vesin
Rosetta Frisone
Gilda Mori Breda
Maria Destefanis
Tiziana Tersilla Comi
Paolina Quarra
Ireneo Zavagna
Carlo Bersano
Elisa Viglezio*
Ugo Cassina
Agostino Borio
Clementina Ferrero*
Piera Chinaglia*
Fausta Audisio
Cesarina Boccalatte
Maria Cibrario

* Names marked with an asterisk are those of Peano's university assistants. There was one other assistant, Ernesto Laura, not included in this list.

LIST OF PAPERS BY OTHER AUTHORS PRESENTED BY PEANO TO THE ACADEMY OF SCIENCES OF TURIN

1892
F. Giudice, 'Sulla risolvente di Malfatti' (1 May).

1893
F. Giudice, 'Sulla risoluzione algebrica delle equazioni' (5 February).
F. Giudice, 'Sulla soluzione dell'equazione algebrica di 5° grado con l'aggiunta della irrazionalità icosaedrale' (7 May).
F. Giudice, 'Sulla determinazione dei numeri reali mediante somme e prodotti' (31 December).

1894
F. Castellano, 'Il complesso delle accelerazioni d'ordine qualunque di punti di un corpo in movimento' (28 January).
C. Burali-Forti, 'Sulle classi derivate a destra e a sinistra' (25 February).

1895
C. Burali-Forti, 'Sul limite delle classi variabili' (20 January).
M. Pieri, 'Sui principii che reggono la geometria di posizione' (19 May).
F. Giudice, 'Sulle equazioni di quinto grado' (17 November).

1896
M. Pieri, 'Sui principii che reggono la geometria di posizione' [2nd note] (26 January).
M. Pieri, 'Sui principii che reggono la geometria di posizione' [3rd note] (9 February).
M. Chini, 'Sulle equazioni a derivate parziali del secondo ordine' (8 March).
C. Burali-Forti, 'Le classi finite' (15 November).
C. Burali-Forti, 'Sopra un teorema del sig. G. Cantor' (13 December).

1897
M. Pieri, 'Sugli enti primitivi della Geometria proiettiva astratta' (24 January).
R. Bettazzi, 'Sulla definizione dei gruppi finiti' (24 January).
A. Ramorino, 'Sopra alcune proprietà delle curve nello spazio in relazione con la loro curvatura a torsione' (7 March).
E. Lampe, 'Sur quelques erreurs dans les *Nuove tavole delle funzioni iperboliche di M. A. Forti*' (21 March).

1898
R. Bettazzi, 'Sulle serie a termini positive, le cui parti rappresentano un continuo' (30 January).
O. Niccoletti, 'Sulle condizioni iniziali che determinano gli integrali delle equazioni differenziali ordinarie" (15 May).
M. Chini, 'Sopra alcune equazioni differenziali" (13 November).

1899

F. Giudice, 'Sull'angolo di due rette e di due piani. Perpendicularità e parallelismo in coordinate omogenee' (15 January).

M. Lerch, 'Nouvelles formules pour la différentiation d'une certaine classe de séries trigonometriques' (19 November).

F. Giudice, 'Sulla metrica degli spazi a curvatura costante' (17 December).

1901

M. Pieri, 'Sui principii che reggono la geometria delle rette' (13 January).

A. Vaccaro, 'Integrazione di sistemi di equazioni lineari differenziali' (14 April).

C. Burali-Forti, 'Sopra alcuni punti singolari delle curve piane e gobbe' (16 June).

1902

C. Burali-Forti, 'Le formole di Frenet per le superfici' (26 January).

C. Burali-Forti, 'Ingranaggi piani' (13 April).

1903

C. Burali-Forti, 'Sul moto di un corpo rigido' (11 January).

T. Boggio, 'Sullo sviluppo in serie di alcune funzioni trascendenti elementari' (11 January).

F. Giudice, 'Sulla integrazione per sostituzione' (14 June).

G. Vitali, 'Sopra le serie di funzioni analitiche' (22 November).

1904

C. Burali-Forti, 'Sulla teoria generale delle grandezze e dei numeri' (3 January).

M. Pieri, 'Circa il teorema fondamentale di Staudt e i principii della geometria proiettiva' (17 January).

M. Chini, 'Sopra una particolare equazione differenziale del $1°$ ordine" (20 November).

F. Giudice, 'Methodo di Newton perfezionato e nuovo metodo pel calcolo assintotico delle radici reali d'equazioni' (4 December).

F. Rimondini, 'Sul calcolo approssimato degli integrali doppi a limiti costanti' (18 (December).

1905

M. Pieri, 'Nuovi principii di Geometria projettiva complessa' (8 January).

F. Castellano, 'Il birapporto di quattro punti nello spazio con applicazioni alla Geometria del Tetraedro' (2 April).

1906

M. Pieri, 'Breve aggiunta alla Memoria *Nuovi principii di Geometria projettiva complessa*' (14 January).

F. Rimondini, 'Sugli integrali definiti di un campo convesso' (29 April).

E. Pascal, 'Sopra una proprietà dei determinanti Wronskiani' (17 June).

A. Padoa, 'Che cosa è una relazione?' (17 June).

C. Burali-Forti, 'Sopra alcune operazioni proiettive applicabili nella meccanica' (18 November).

1907

C. Burali-Forti, 'Sulle omografie vettoriali' (10 February).

C. Burali-Forti, 'Funzioni vettoriali' (17 November).

F. Giudice, 'Una dimostrazione d'inseparabilità per radicali delle 27 rette di superficie cubiche' (1 December).

1908

C. Burali-Forti, 'I quaternioni di Hamilton e il calcolo vettoriale' (17 June).

T. Boggio, 'Sulla risoluzione di una classe di equazioni algebriche che si presentano nella matematica finanziaria e attuariale' (15 November).

1909

G. Pagliero, 'Geodetica di una superficie di rivoluzione' (9 May).

M. Peyroleri, 'Relazione fra calcolo delle differenze e calcolo differenziale' (13 June).

F. Giudice, 'Sull'inscrivibilità circolare dei poligoni articolati' (13 June).

C. Burali-Forti, 'Sulla Geometria differenziale assoluta delle congruenze e dei complessi rettilinei' (21 November).

C. A. Dell'Agnola, 'Sul teorema di Borel' (21 November).

1910

T. Boggio, 'Dimostrazione assoluta delle equazioni classiche dell'Idrodinamica' (30 January).

C. Burali-Forti, 'Gradiente, rotazione e divergenza in una superficie' (27 February).

G. Santangelo, 'Su di una estensione del teorema di Habich' (27 February).

M. Gramegna, 'Serie di equazioni differenziali lineari ed equazioni integro-differenziali' (13 March).

1911

G. Pagliero, 'Resto nella formula di Lubbock' (19 February).

C. Burali-Forti, 'Alcune applicazioni alla geometria differenziale su di una superficie dell'operatore omografico C' (5 March).

C. Burali-Forti, 'Sopra una formula generale per la trasformazione di integrali di omografie vettoriali' (21 May).

G. Pagliero, 'I numeri primi da 100.000.000 a 1000.005.000' (21 May).

M. Bottasso, 'Alcune applicazioni alle formule di Frenet' (19 November).

C. Burali-Forti, 'Sul moto composto' (31 December).

1912

A. Pastore, 'Le definizioni matematiche secondo Aristotele e la Logica matematica' (10 March).

A. Padoa, Frequenza, previsione, probabilità' (26 May).

M. Bottasso, 'Il teorema di Rouché-Capelli per i sistemi di equazioni integrali' (17 November).

T. Astuti, 'Sulla trasformazione di Tschirnhausen' (17 November).

M. Pieri, 'Sui sistemi di ∞^1 superficie' (1 December).

A. Pensa, 'Sopra alcuni operatori differenziali omografici' (1 December).

1913

F. Giudice, 'Interpretazione geometrica del Metodo di Lagrange' (26 January).

P. Quarra, 'Resto in alcune formule di quadratura' (30 March).

V. Mago, 'Teoria degli ordini' (27 April).

L. Tonelli, 'Sul valore di un certo ragionamento' (16 November).

C. Poli, 'Sulla dimostrazione della integrabilità delle funzioni continue' (30 November).

M. Bottasso, 'Sulle determinazione del tasso di una rendita temporanea, variabile e continua' (28 December).

1914

M. Bottasso, 'Sopra alcune estensioni dei teoremi di Guldino' (22 February).

A. Pensa, 'Sulla risoluzione di equazioni vettoriali ed omografiche' (10 May).

A. Pensa, 'Alcune applicazioni delle formule di Frenet' (14 June).

S. A. Toscano, 'Sopra un inviluppo di circonferenze' (29 November).

C. Burali-Forti, 'Isomerie vettoriali e Moti geometrici' (13 December).

1915

T. Boggio, 'Sul problema delle vene confluenti' (7 March).

M. Bottasso, 'Sopra un nuovo problema dei valori al contorno per un cerchio' (7 March).

C. Burali-Forti, 'Nuove applicazioni degli operatori' (7 March).

F. Castellano, 'I numeri complessi considerati come operatori sui vettori di un piano' (21 March).

G. Sannia, 'I limiti di una funzione in un punto limite del suo campo" (11 April).

A. Tanturri, 'Prodotto di due numeri approssimati. Errore relativo o errore assoluto?' (25 April).

G. Andreoli, 'Su un problema di meccanica ereditaria" (9 May).

S. Catania, "Sulle condizioni che caratterizzano una classe di grandezze" (14 November).

F. Cantelli, 'Resti nelle formole di quadratura' (28 November).

1916

G. Vacca, 'Sul poligono regolare di 17 lati' (6 February).

M. Bottasso, 'Teoremi su massimi e minimi geometrici; e su normali a curve e superficie' (2 April).

A. Tanturri, 'Radici di numeri approssimati ed estrazione abbreviata della radice quadrata' (21 May).

L. Della Casa, 'Rapporto di grandezze eterogenee' (21 May).

L. Viriglio, 'I segni numerali romani' (19 November).

A. Padoa, 'Corrispondenze duali' (19 November) [never printed].

1917

C. Burali-Forti, 'Equivalenti omografiche delle Formole di Frenet – Linee e superficie parallele' (13 May).

R. Frisone, 'Una teoria semplice dei logaritmi' (13 May).

A. Tanturri, 'Della partizioni dei numeri. Ambi, terni, quaterne e cinquine di data somma' (27 May).

A. Pensa, 'Sull'operatore omografico R'' (2 December).

C. Burali-Forti, 'Alcuni sistemi di linee su di una superficie' (16 December).

E. Boverio, 'Sopra la derivazione dei canali' (16 December).

1918

G. Mori Breda, 'Estrazione della radice quadrata' (13 January).

C. Burali-Forti, 'Linea in ogni cui punto è assegnata una direzione invariabilmente collegata al triedro principale' (24 February).

R. Frisone, 'Le varie definizioni di prodotto' (10 March).

Lagneau, 'Logique des propositions' (10 March).

M. Bottasso, 'Generalizzazione della trasformazione di Combescure per le curve (28 April).

A. Tanturri, 'Sui prodotti infiniti $(1 - x)(1 - x^2)(1 - x^3) \ldots$ e $(1 + x)(1 + x^2)(1 + x^3) \ldots$' (12 May).

P. Quarra, 'Calcolo delle parentesi' (16 June).

L. Viriglio, 'Estrazione graduale di radice cubica' (16 June).

T. T. Comi, 'Formule sommatorie' (17 November).

A. Tanturri, 'Sul numero delle partizioni di un numero in potenze di 2' (1 December).

T. Boggio, 'Sulla Geometria assoluta degli spazi curvi' (15 December).

O. Lazzarino, 'Sulla generalizzazione dei moti alla Poinsot e sulla stabilità degli assi permanenti di rotazione in detti moti' (15 December).

1919

A. Tanturri, 'Sulla funzione di Dirichlet e sulla funzione signum x del Kronecker' (23 February).

G. Mori Breda, 'Sviluppo delle radici in prodotto decimale' (23 March).

T. T. Comi, 'Sviluppo delle radici in prodotto decimale' (23 March).

I. Zavagna, 'Calcolo dei logaritmi naturali con la serie esponenziale' (15 June).

C. Bersano, 'Il numero π calcolato con la serie esponenziale' (15 June).

F. Sibirani, 'Espressioni analitiche che definiscono più funzioni analitiche ad area lacunare' (30 November).

1921

A. Tanturri, 'Saggio di rappresentazioni analitiche di funzioni singolari' (17 April).

E. Viglezio, 'Aree di curve piane' (29 May).

1922

U. Cassina, 'Volume, area, lunghezza e curvatura di una figura' (22 January).

C. Burali-Forti, 'Operatori per le iperomografie' (19 February).

E. Viglezio, 'Calcolo diretto dei logaritmi decimali' (31 December).

1923

U. Cassina, 'Risoluzione graduale dell'equazione cubica di Leonardo Pisano' (18 November).

F. Sibirani, 'Sulla sfera paraosculatrice ad una curva storta' (18 November).

1924

C. Ferrero, 'Resto nella Formula di quadratura Cavalieri-Simpson' (28 December).

1925

C. Burali-Forti, 'Stato cinetico, moto infinitesimo, teorema di Coriolis' (11 January).

P. Vocca, 'Studio sulla latitudine della 1^a Sala meridiana del R. Osservatorio astronomico di Pino Torinese' (24 May).

1926
U. Cassina, 'Limiti delle funzioni plurivoche' (14 November).

1928
U. Cassina, 'Nuova teorie delle grandezze' (26 February).
G. Vacca, 'Sul principio della descesa di Fermat e sulle dimostrazioni dell'esistenza degli irrazionali quadratici' (29 April).
C. Boccalatte, 'La Geometria basata sulle idee di punto e angolo retto' (2 December).

1929
V. Marseguerra, 'Sulle funzioni tayloriane' (5 May).
M. Cibrario, Proposizioni universali e particolari e definizione del limite' (23 June).

1930
F. Audisio, 'Calcolo di π in Archimede' (2 February).

1931
U. Cassina, Grave in terra rotante' (14 June).
M. Gliozzi, 'Precursori del sistema metrico decimale' (15 November).

1932
G. Arrighi, 'Introduzione alla geometria differenziale della superficie dei centri di carena' (6 March).

CHRONOLOGICAL LIST OF THE PUBLICATIONS OF GIUSEPPE PEANO

In the chronological list published in the three-volume *Opere Scelte*, Ugo Cassina used the consecutive numbering of Peano's publications which had already been used by Peano himself. He also distinguished between mathematical and philological publications, leaving only one publication outside these two categories (and outside the consecutive numbering.) Since the *Opere Scelte* will be the most available source, for most readers, of the original articles, I have kept the numbering of Cassina and have used a decimal numbering to insert new titles into his list. This new list was published in *Selected Works of Giuseppe Peano*; to it has been added only a few translations that have since come to my attention.

I have followed Cassina's use of Roman numerals to distinguish monographs and volumes (of at least 64 pages), primes to denote translations and reprints, and the asterisk to signalize works pertaining to philology. Thus *160'−17' is a reprint of the 160th work and the 17th work in philology. Titles of journals are those of *Union List of Serials*; abbreviations of journal titles are those of *Mathematical Reviews* (with the addition of *A.p.I.* and *R.d.M.*). In cases where the original article was untitled, I have given a brief descriptive phrase in English. This suffices to show that the article was untitled since none of Peano's publications was in English.

The following abbreviations of journal titles are used:

A.p.I.   Academia pro Interlingua
Amer. J. Math.   American Journal of Mathematics
Ann. mat. pura appl.   Annali di matematica pura ed applicata
Atti Accad. naz. Lincei, Rend., Cl. sci. fis. mat. nat.   Atti della Accademia Nazionale dei Lincei, Rendiconti, Classe di scienze fisiche, matematiche e naturali
Atti Accad. sci. Torino, Cl. sci. fis. mat. nat.   Atti della Accademia delle scienze di Torino, Classe di scienze fisiche, matematiche e naturali
Boll. Un. mat. ital.   Bollettino della Unione matematica italiana
Enseignement math.   L'Enseignement mathématique
Giorn. mat. Battaglini   Giornale di matematiche di Battaglini
Giorn. mat. finanz.   Giornale di matematica finanziaria: Rivista tecnica del credito e della previdenza

195

Math. Ann.   Mathematische Annalen
Mathesis   Mathesis: Recueil mathématique à l'usage des écoles spéciales et des établissements d'instruction moyenne
Monatsh. Math.   Monatshefte für Mathematik
Period. mat.   Periodico di matematiche
R.d.M.   Rivista di matematica (Revue de mathématiques)
Rend. Circ. mat. Palermo   Rendiconti del Circolo matematico di Palermo
Scientia   Scientia: International Review of Scientific Synthesis
Wiadom. mat.   Rocziki Polskiego towarzystwa matematycznego, ser. II: Wiadomości matematyczne

Peano, Su no Gainen ni tsuite (Kyoritsu Publishing Company, 1969) has Japanese translations of [35] and [37] by T. Umezawa, with a commentary by K. Ono. *Selected Works of Giuseppe Peano*, H. C. Kennedy editor-translator (University of Toronto Press, 1973), has English translations of items: 5, 9, 10, 16, 22, 23, 24, 29, 31, 50, 62, 90, 90.5, 91, 133′ (Additione), 143, 176, 193, and parts of 8, 11, 14.

1881

1 'Construzione dei connessi (1, 2) e (2, 2)', *Atti Accad. sci. Torino* **16**, 497−503.
2 'Un teorema sulle forme multiple', *Atti Accad. sci. Torino* **17**, 73−9.
3 'Formazioni invariantive delle corrispondenze', *Giorn. mat. Battaglini* **20**, 79−100.

1882

4 'Sui sistemi di forme binarie di egual grado, e sistema completo di quante si vogliano cubiche', *Atti Acad. sci. Torino* **17**, 580−6.

1883

5 'Sull'integrabilità delle funzioni', *Atti Accad. sci. Torino* **18**, 439−46.
6 'Sulle funzioni interpolari', *Atti Accad. sci. Torino* **18**, 573−80.

1884

7 Two letters to the Editor, *Nouvelles annales de mathématiques* (3) **3**, 45−7, 252−6.
I, 8 Angelo Genocchi, *Calcolo differenziale e principii di calcolo integrale, pubblicato con aggiunte dal Dr. Giuseppe Peano* (Torino: Bocca).
I′, 8′ Angelo Gnocchi, *Differentialrechnung und Grundzüge der Integralrechnung, herausgegeben von Giuseppe Peano*, trans. by G. Bohlmann and A. Schepp, with a preface by A. Mayer (Leipzig: Teubner, 1899).
Russian translations by N. S. Sineokov (Kiev, 1903) and K. A. Posse (St. Petersburg, 1922).

1886

9 'Sull'integrabilità delle equazioni differenziali del primo ordine', *Atti Accad. sci. Torino* **21**, 677–85.

1887

10 'Integrazione per serie delle equazioni differenziali lineari', *Atti Accad. sci. Torino* **22**, 437–46.

II, 11 *Applicazioni geometriche del calcolo infinitesimale* (Torino: Bocca).

1888

12 'Intégration par séries des équations différentielles linéaires', *Math. Ann.* **32**, 450–6 (English translation in [Birkhoff 1973]).

13 'Definizione geometrica delle funzioni ellittiche', *Giom. mat. Battaglini* **26**, 255–6.

13′ 'Definição geometrica das funcções ellipticas', *Jornal de sciencias mathematicas e astronomicas* **9**, 24–5.

III, 14 *Calcolo geometrico secondo l'Ausdehnungslehre di H. Grassmann, preceduto dalle operazioni della logica deduttiva* (Torino: Bocca).

14.1 Communications to the Circolo (on a note of F. Giudice and on his response), *Rend. Circ. mat. Palermo* **2**, 94, 187–8.

15 'Teoremi su massini e minimi geometrici, e su normali a curve e superficie', *Rend. Circ. mat. Palermo* **2**, 189–92.

1889

16 *Arithmetices principia, nova methodo exposita* (Torino: Bocca).

17 'Sur les wronskiens', *Mathesis* **9**, 75–6, 110–12.

18 *I principii di geometria logicamente esposti* (Torino: Bocca).

19 'Une nouvelle forme du reste dans la formule de Taylor', *Mathesis* **9**, 182–3.

20 'Su d'una proposizione riferentesi ai determinanti jacobiani', *Giorn. mat. Battaglini* **27**, 226–8.

21 Angelo Genocchi, *Annuario R. Università di Torino* (1889–90), 195–202.

22 'Sur une formule d'approximation pour la rectification de l'ellipse', *Académie des sciences, Paris, Comptes-rendus hebdomadaire des séances* **109**, 960–1.

1890

23 'Sulla definizione dell'area d'una superficie', *Atti Accad. naz. Lincei, Rend., Cl. sic. fis. mat. nat.* (4) **6–I**, 54–7.

24 'Sur une courbe qui remplit toute une aire plane', *Math. Ann.* **36**, 157–60.

25 'Les propositions du V livre d'Euclide réduites en formules', *Mathesis* **10**, 73–5.

26 'Sur l'interversion des dérivations partielles', *Mathesis* **10**, 153–4.

27 'Démonstration de l'intégrabilité des équations différentielles ordinaires', *Math. Ann.* **37**, 182–228.

28 'Valori approssimati per l'area di un ellissoide', *Atti Accad. naz. Lincei, Rend., Cl. sci. fis. mat. nat.* (4) 6–II, 317–21.
29 'Sopra alcune curve singolari', *Atti Accad. sci. Torino* 26, 299–302.

1891

30 *Gli elementi di calcolo geometrico* (Torino: Candeletti).
30′ *Die Grundzüge des geometrischen Calculs*, trans. by A. Schepp (Leipzig: Teubner).
31 'Principii di logica matematica', *R.d.M.* 1, 1–10.
31′ 'Principios de lógica matemática', *El progreso matemático* 2 (1892), 20–4, 49–53.
32 'Sommario dei libri VII, VIII, IX d'Euclide', *R.d.M.* 1, 10–12.
33 Review of E. W. Hyde, *The Directional Calculus, based on the methods of H. Grassmann*, *R.d.M.* 1, 17–19.
34 Review of F. D'Arcais, *Corso di calcolo infinitesimale*, *R.d.M.* 1, 19–21.
35 'Formule di logica matematica', *R.d.M.* 1, 24–31, 182–4.
36 Observations on an article of C. Segre, *R.d.M.* 1, 66–9.
37 'Sul concetto di numero', *R.d.M.* 1, 87–102, 256–67.
38 Review of S. Dickstein, *Projecia i metody matematyki*, *R.d.M.* 1, 124,
38.1 Reply to a declaration of C. Segre, *R.d.M.* 1, 156–9.
39 Review of E. Schröder *Vorlesungen über die Algebra der Logik*, *R.d.M.* 1, 164–70.
40 Open letter to Prof. G. Veronese, *R.d.M.* 1, 267–9.
41 'Il teorema fondamentale di trigonometria sferica', *R.d.M.* 1, 269.
42 'Sulla formula di Taylor', *Atti Accad. sci. Torino* 27, 40–6.
42′ 'Ueber die Taylor'sche Formel', Anhang III in [I′, 8′], 359–65.
42.1 'Questions proposées, no. 1599', *Nouvelles annales de mathématiques* (3) 10, 2*.
a   *Rivista di matematica*, vol. 1 (Torino: Bocca).

1892

43 Comments on an unsolved problem, *R.d.M.* 2, 1–2.
44 'Sommario del libro X d'Euclide', *R.d.M.* 2, 7–11.
44.1 (With F. Giudice) Review of Domenico Amanzio, *Elementi di algebra elementare*, *R.d.M.* 2, 14–17.
45 'Sur la définition de la dérivée', *Mathesis* (2) 2, 12–14.
46 'Osservazioni sul 'Traité d'analyse par H. Laurent'', *R.d.M.* 2, 31–4.
46.1 (Unsigned) 'Enrico Novarese', *R.d.M.* 2, 35.
47 'Esempi di funzioni sempre crescenti e discontinue in ogni intervallo', *R.d.M.* 2, 41–2.
48 Proposed question no. VI, *R.d.M.* 2, 42.
49 'Sur le théorème général relatif à l'existence des intégrales des équations différentielles ordinaires', *Nouvelles annales de mathématiques* (3) 11, 79–82.
50 'Generalizzazione della formula di Simpson', *Atti Accad. sci. Torino* 27, 608–12.

51 'Dimostrazione dell'impossibilità di segmenti infinitesimi costanti', *R.d.M.* **2**, 58–62.
52 'Sulla definizione del limite d'una funzione', *R.d.M.* **2**, 77–9.
53 Review of Albino Nagy, *Lo stato attuale ed i progressi della logica*, *R.d.M.* **2**, 80.
54 A brief reply to Prof. Veronese, *Rend. Circ mat. Palermo* **6**, 160.
55 Letter to the Editor (Extrait d'une lettre de M. Peano à M. Brisse), *Nouvelles annales de mathématiques* (3) **11**, 289.
56 Review of G. Veronese, *Fondamenti di geometria a più dimensioni*, etc., *R.d.M.* **2**, 143–4.
b *Rivista di matematica*, vol. 2 (Torino: Bocca).

1893

57 (With E. D'Ovidio and C. Segre) 'Relazione intorno alla memoria del Prof. V. Mollame 'Sulle equazioni abeliane reciproche, le cui radici si possono rappresentare on con x, θx, $\theta^2 x$, ... $\theta^{n-1}x$', *Atti Accad. sci. Torino* **28**, 781.
58 'Formule di quadratura', *R.d.M.* **3**, 17–18.
58.1 Review of N. Jadanza, *Una difesa delle formole di Simpson, ed alcune formole di quadratura poco note*, *R.d.M.* **3**, 137.
59 Review of E. Carvallo, *Sur les forces centrales*, *R.d.M.* **3**, 137–8.
IV, 60 *Lezioni di analisi infinitesimale* (2 vols.; Torino: Candeletti).
60' 'Die komplexen Zahlen', (a translation of chapter 6), Anhang V in [I', 8'], 371–95.
61 Review of A. Ziwet, *An Elementary Treatise on Theoretical Mechanics*, Part I: *Kinematics*, *R.d.M.* **3**, 184.
c *Rivista di matematica*, vol. 3 (Torino: Bocca).

1894

62 'Sur les systèmes linéaires', *Monatsh. Math.* **5**, 136.
63 'Sulla parte V del Formulario: Teorie dei gruppi di punti', *R.d.M.* **4**, 33–5.
64 'Sui fondamenti della geometria', *R.d.M.* **4**, 51–90.
64.1 A brief commemoration of E. Catalan, *R.d.M.* **4**, 104.
65 'Un precursore della logica matematica', *R.d.M.* **4**, 120.
66 *Notations de logique mathématique* (Introduction au formulaire de mathématiques) (Turin: Guadagnini).
67 'Notions de logique mathématique', *Association française pour l'avancement des sciences, 23. session, Caen, 1894, Compte rendu* (Paris, 1895), 222–6.
68 'Sur la définition de la limite d'une fonction. Exercise de logique mathématique', *Amer. J. Math.* **17**, 27–68.
69 Review of Hermann Grassmann's *Gesammeltemathematische und physikalische Werke*, vol. I, part I, *R.d.M.* **4**, 167–9.
70 'Estensione di alcuni teoremi di Cauchy sui limiti', *Atti Accad. sci. Torino* **30**, 20–41.
70.1 Reply to question no. 14, *L'intermédiaire des mathématiciens* **1**, 157–8.

70.2 Reply to question no. 102, *L'intermédiaire des mathématiciens* 1, 195–6.

70.3 (With G. Vivanti) *Teoria dei gruppi di punti* (Torino: Fodratti e Lecco).
 d *Rivista di matematica*, vol. 4 (Torino: Bocca).

1895

V, 71 *Formulaire de mathématiques*, vol. 1 (Torino: Bocca).
 72 (With G. Vailati) 'Logique mathématique', [V, 71], 1–8.
 73 (With F. Castellano) 'Opérations algébriques', [V, 71], 8–22.
 74 (With C. Burali-Forti) 'Arithmétique', [V, 71], 22–8.
 75 'Classes de nombres', [V, 71], 58–65.
 76 Review of F. Castellano, *Lezioni di meccanica razionale*, *R.d.M.* 5, 11–18.
 77 'Il principio delle aree e la storia di un gatto', *R.d.M.* 5, 31–2.
 78 'Sulla definizione di integrale', *Ann. mat. pura appl.* (2) 23, 153–7.
 78' 'Ueber die Definition des Integrals', Anhang IV in [I', 8'], 366–70.
 79 'Sopra lo spostamento del polo sulla terra', *Atti Accad. sci. Torino* 30, 515–23.
 80 'Sul moto del polo terrestre', *Atti Accad. sci. Torino* 30, 845–52.
 81 Letter to the Editor, *Monatsh. Math.* 6, 204.
 82 Review of G. Frege, *Grundgesetze der Arithmetik, begriffsschriftlich abgeleitet*, vol. 1, *R.d.M.* 5, 122–8 (English translation in [Dudman 1971]).
 83 'Elenco bibliografico sull'"Ausdehnungslehre" di H. Grassmann', *R.d.M.* 5, 179–82.
 84 'Sul moto d'un sistema nel quale sussistono moti interni variabili', *Atti Accad. naz. Lincei, Rend., Cl. sci. fis. mat. nat.* (5) 4–II, 280–2.
 85 'Trasformazioni lineari dei vettori di un piano', *Atti Accad. sci. Torino* 31, 157–66.
 86 (With E. D'Ovidio) 'Relazione sulla memoria del Prof. Francesco Giudice, intitolata: 'Sull'equazione del 5° grado'', *Atti Accad. sci. Torino* 31, 199.
86.1 Reply to question no. 288, *L'intermédiaire des mathématiciens* 2, 83.
 e *Rivista di matematica*, vol. 5 (Torino: Bocca).

1896

 87 'Sul pendolo di lunghezza variabile', *Rend. Circ. mat. Palermo* 10, 36–7.
 88 'Introduction au tome II du *Formulaire de mathématiques*', *R.d.M.* 6, 1–4.
 89 'Sul moto del polo terrestre', *Atti Accad. naz. Lincei, Rend., Cl. sci. fis. mat. nat.* (5) 5–I 163–8.
 90 'Saggio di calcolo geometrico', *Atti Accad. sci. Torino* 31, 952–75.
 90' *Zarys Rachunku geometrycznego*, trans. by S. Dickstein (Warsaw, 1897).
 90" *Entwicklung der Grundbegriffe des geometrischen Calculs*, trans. by A. Lanner (Salzburg, 1897–8).

90.1 Reply to question no. 60, *L'intermédiaire des mathématiciens* 3, 39.
90.2 Reply to question no. 101, *L'intermédiaire des mathématiciens* 3, 61.
90.3 Reply to question no. 362, *L'intermédiaire des mathématiciens* 3, 69.
90.4 Reply to question no. 80, *L'intermédiaire des mathématiciens* 3, 87.
90.5 Reply to question no. 719, *L'intermédiaire des mathématiciens* 3, 169.

1897

91 'Studii di logica matematica', *Atti Accad. sci. Torino* 32, 565–83.
91' 'Ueber mathematische Logik', Anhang I in [I', 8'], 336–52.
92 'Sul determinante wronskiano', *Atti Accad. naz. Lincei, Rend., Cl. sci. fis. mat. nat.* (5) 6–I, 413–15.
VI, 93 *Formulaire de mathématiques*, vol. 2, section 1: *Logique mathématique* (Turin: Bocca).
93' 'Logica matematica', *Verhandlungen des ersten internationalen Mathematiker-Kongresses, in Zürich vom 9. bis 11. August 1897* (Leipzig, 1898).
94 'Generalità sulle equazioni differenziali ordinaire', *Atti Accad. sci. Torino* 33, 9–18.
95 (With E. D'Ovidio and C. Segre) 'Relazione sulla memoria 'I principii della geometria di posizione composti in sistema logico-deduttivo' del Prof. M. Pieri', *Atti Accad. sci. Torino* 33, 148–50.

1898

96 'Sulle formule di logica', *R.d.M.* 6, 48–52.
97 Reply to a letter of G. Frege, *R.d.M.* 6, 60–1 (English translation in [Dudman 1971]).
98 'Analisi della teoria dei vettori', *Atti Accad. sci. Torino* 33, 513–34.
VII, 99 *Formulaire de mathématiques*, vol. 2, section 2 (Turin: Bocca).
99' Definitionen der Arithmetik', Anhang II in [I', 8'], 353–8.
100 (With others) 'Additions et corrections à F_2', *R.d.M.* 6, 65–74.
101 'Sul § 2 del Formulario t. II: Aritmetica', *R.d.M.* 6, 75–89.
102 Review of E. Schröder, Ueber Pasigraphie etc., *R.d.M.* 6, 95–101.
103 'La numerazione binaria applicata alla stenografia', *Atti Accad. sci. Torino* 34, 47–55.
103.1 Reply to question no. 1132, *L'intermédiaire des mathématiciens* 5, 23.
103.2 Reply to question no. 1185, *L'intermédiaire des mathématiciens* 5, 71–2.
103.3 Question no. 1297, *L'intermédiaire des mathématiciens* 5, 125–6.
103.4 Reply to question no. 1208, *L'intermédiaire des mathématiciens* 5, 144.

1899

104 Letter to the Editor, *Period. mat.* (2) 14, 152–3.
105 'Sui numeri irrazionali', *R.d.M.* 6, 126–40.
VIII, 106 *Formulaire de mathématiques*, vol. 2, section 3 (Turin: Bocca).

106.1 Reply to question no 1374, *L'intermédiaire des mathématiciens* 6, 135.

   f *Revue de mathématiques*, vol. 6 (Turin: Bocca, 1896–9).

1900

107 'Formules de logique mathématique', *R.d.M.* 7, 1–41.

108 'Additions au Formulaire', *R.d.M.* 7, 67–70.

109 'Les définitions mathématiques', *Congrès international de philosophie, Paris, 1900* (Paris, 1901), vol. 3, 279–88.

109′ 'Definicye w matematyce', trans. by Z. Krygowski, *Wiadom. mat.* 6, 174–81.

1901

IX, 110 *Formulaire de mathématiques*, vol. 3 (Paris: Carré and Naud).

111 *Studio delle basi sociali della Cassa nazionale mutua cooperativa per le pensioni* (Torino).

111′ *Studio delle basi sociali della Cassa mutua cooperativa per le pensioni*, 2nd ed. (Torino, 1906).

112 (With others) 'Additions et corrections au Formulaire a. 1901', *R.d.M.* 7, 85–110.

113 Review of O. Stolz and I. A. Gmeiner, *Theoretische Arithmetik*, I, *R.d.M.* 7, 112–14.

114 *Relazione della Commissione per lo studio delle basi sociali della Cassa etc.* (Torino).

115 'Memoria', *Bollettino di notizie sul credito etc., del Ministero agr. ind. e comm.* (1901), 1200–(?).

116 *Dizionario di logica matematica*, Presented to the Congress of Secondary School Professors, Livorno, 1901.

116′ 'Dizionario di matematica, parte I, Logica matematica', *R.d.M.* 7, 160–72.

116″ 'Dizionario di matematica, parte I, Logica matematica', (Torino).

116.1 Letter to the Editor, *Il bollettino di matematiche e di scienze fisiche e naturali. Giornale per la coltura dei maestri delle scuole elementari e degli alunni delle scuole normale* 2, 254–5.

117 (With others) 'Additions au Formulaire a. 1901', *R.d.M.* 7, 173–84.

   g *Revue de mathématiques*, vol. 7 (Turin: Bocca, 1900–1).

1902

X, 118 *Aritmetica generale e algebra elementare* (Torino: Paravia).

119 Review of C. Arzelà, *Lezioni di cacolo infinitesimale*, vol. I, part 1, *R.d.M.* 8, 7–11.

120 *Seconda relazione della Commissione per lo studio delle basi sociali della Cassa, etc.* (Torino).

121 Letter to the Editor (Sur les imaginaires), *Bulletin des sciences mathématiques et physiques élémentaires* 7, 275–7.

122 'La geometria basata sulle idee di punto e distanza', *Atti Accad. sci. Torino* 38, 6–10.

123 *A proposito di alcuni errori contenuti nel disegno di legge sulle asso-*
     ciazioni tontinarie presentato al Senato (Torino).

1903

124 *Sul massimo della pensione a distribuirsi dalla Cassa etc.* (Torino).

XI, 125 *Formulaire mathématique*, vol. 4 (Turin: Bocca).

*126–1 'De latino sine-flexione', *R.d.M.* 8, 74–83.

127–126 'Principio de permanentia', *R.d.M.* 8, 84–7.

*127'–1' *De latino sine flexione – Principio de permanentia* (Torino).

127.1 'Question no. 2549', *L'intermédiaire des mathématiciens* 10, 70.

1904

*128–2 'Il latino quale lingua ausiliare internazionale', *Atti Accad. sci. Torino*
        39, 273–83.

129–127 'Sur les principes de la géométrie selon M. Pieri', *Kazan Universitet,*
         Fiziko-matematicheskoe obshchestvo, Izvestiya (2) 14 (1905),
         92–5.

*130–3 *Vocabulario de latino internationale, comparato cum Anglo, Franco,*
        Germano, Hispano, Italo, Russo, Graeco et Sanscrito (Torino).

1905

131–128 *Progetto di una Cassa di riassicurazione e di una Cassa di soccorso etc.*
         (Torino).

131.1 (With C. Segre) 'Relazione sulla memoria del prof. Mario Pieri: Nuovi
       principî di geometria projettiva complessa', *Atti Accad. sci. Torino*
       40, 378–9.

1906

132–129 'Sulle differenze finite', *Atti Accad. naz. Lincei, Rend., Cl. sci. fis. mat.*
         nat. (5) 15–I, 71–2.

133–130 'Super theorema de Cantor-Bernstein', *Rend. Circ. mat. Palermo* 21,
         360–6.

133'–130' 'Super theorema de Cantor-Bernstein et additione', *R.d.M.* 8, 136–57.

*134–4 'Notitias super lingua internationale', *R.d.M.* 8, 159.

     h *Revista de mathematica*, vol. 8 (Torino: Bocca, 1902–6).

135–131 'Sul libro V di Euclide', *Bollettino di matematica* 5, 87–91.

136–132 'Sur la convergence absolue des séries' et 'sur un développement en série
         entière', *Enseignement math.* 8, 315–6.

*137–5 *Formulario mathematico*, ed. V, *Indice et vocabulario* (Torino: Bocca).

137.1 Reply to question no. 854, *L'intermédiaire des mathématiciens* 13,
       39.

1907

137.2 Letter to S. Catania, *Il Pitagora* 13, 92–4.

1908

XII, 138–133 *Formulario mathematico*, vol. 5 (Torino: Bocca, 1905–8).

204   GIUSEPPE PEANO – LIST OF PUBLICATIONS

*139–6 'Mensura de internationalitate', *Correspondens international* (A. Miller) (London).
139.1 'Curriculo de vita de G. Peano; Pubblicazioni del Prof. G. Peano', *Akademi internasional de lingu universal, Sirkular* nr. 48 (93), 2–6.

1909

140–134 (With C. Somigliana) 'Relazione intorno alla memoria del Dr. Tommaso Boggio, intitolata: Sulla rizoluzione di una classe di equazioni algebriche che si presentano nella matematica finanziaria e attuariale', *Atti Accad. sci. Torino* 44, 171.
XIII, *141–7 *Vocabulario commune ad linguas de Europa* (Torino: Bocca).
142–135 'Notations rationelles pour le système vectoriel à propos du système proposé par MM. Burali-Forti et Marcolongo. No. 5. Lettre de M. Peano (Turin)', *Enseignement math.* 11, 216–7.
142.1 'In memoria di Giovanni Vailati', *Gazzetta del popolo* (Torino), 17 May 1909. Reprinted in *Bollettino di matematica* 8, 206–7.

1910

143–136 'Sui fondamenti dell'analisi', *Mathesis, società italiana di matematica, Bollettino* 2, 31–7.
144–137 'Sugli ordini degli infiniti', *Atti Accad. naz. Lincei, Rend., Cl. sci. fis. mat. nat.* (5) 19–I, 778–81.
*145–8 'A proposito della lingua internazionale', *Classici e neo-latini* 6, no. 4.
*146–9 (See page numbers for titles) *Academia pro Interlingua, Discussiones* 1, 1–6 (untitled); 9–13, Propositiones in discussione; 57–8, Propositiones in discussione; (with G. Pagliero) 77–81, 'Discussiones' in 'Progreso'.
*147–10 'Lingua de Academia', *A.p.I Discussiones* 1, 91–6, 147–57, 187–91.
*148–11 'Exemplo de Interlingua', *A.p.I. Discussiones* 1, 163–74.
i *Academia pro Interlingua, Discussiones*, vol. 1 (Torino, 1909–10).

1911

149–138 'Sulla definizione di funzione', *Atti Accad. naz. Lincei, Rend., Cl. sci. fis. mat. nat.* (5) 20–I, 3–5.
*150–12 'Le latin sans flexions', *Les questions modernes* 2, 509–12.
151–139 'Le definizioni in matematica', *Institut d'estudis catalans, Barcelona, Seccio de ciencies, Arxius* 1, no. 1, 49–70.
*152–13 'De passivo', *A.p.I. Discussiones* 2, 81–4.
*153–14 *100 exemplo de Interlingua cum vocabulario Interlingua-italiano* (Torino: Bocca).
*153'–14' *100 exemplo de Interlingua cum vocabulario Interlingua-latino-italiano-français-english-deutsch*, 2nd ed. (Torino, 1913).
*154–15 Una questione di grammatica razionale', *Fourth International Congress of Philosophy, Bologna, 5–11 April 1911* (Genoa: Formiggini), vol. 2, 343–8.
154.1 'Observaciones etimológicas', *Revista de la Sociedad matemática epañola* 1, 143.
j *Academia pro Interlingua, Discussiones*, vol. 2 (Torino: Bocca).

1912

*155–16   'De derivatione', *A.p.I. Discussiones* 3, 20–43.

156–140   'Sulla definizione di probabilità', *Atti Accad. naz. Lincei, Rend., Cl. sci. fis. mat. nat.* (5) 21–I, 429–31.

157–141   'Delle proposizioni esistenziali', *Fifth International Congress of Mathematicians, Cambridge, 22–28 August 1912* (Cambridge: University Press, 1913), vol. 2, 497–500.

158–142   'Contro gli esami', *Torino nuova* (17 August 1912), 2.

159–143   'Derivata e differenziale', *Atti. Accad. sci. Torino* 48, 47–69.

159.1   Review of F. Castellano, *Lezioni di meccanica razionale*, seconda edizione, *Bollettino di bibliografia e storia delle scienze matematiche* 14, 7.

159.2   'Preface' to Alessandro Padoa, *La logique déductive dans sa dernière phase de développement* (Paris: Gauthier-Villars).

159.3   (With C. Somigliana) 'Relazione sulla memoria del Dr. G. Sannia: Caratteristiche multiple di un'equazione alle derivate parziali in due variabili indipendenti', *Atti Accad. sci. Torino* 48, 196.

159.4   (Unsigned) 'Martin Schleyer', *A.p.I. Discussiones* 3, 164–70.

k   *Academia pro Interlingua, Discussiones*, vol. 3 (Torino: Bocca).

1913

*160–17   'Vocabulario de Interlingua', *Ap.I. Discussiones* 4, 13–19.

*160'–17'   'De vocabulario', *A.p.I. Circulare* (1924), no. 4, 9–13. Reprinted in *Schola et Vita* 1 (1926), 191–5.

161–144   'Sulla definizione di limite', *Atti Accad. sci. Torino* 48, 750–72.

162–145   'Resto nelle formule di quadratura espresso con un integrale definito', *Atti Accad. naz. Lincei, Rend., Cl. sci. fis. mat. nat.* (5) 22–I, 562–9.

163–146   Review of A. N. Whitehead and B. Russell, *Principia mathematica*, vols. I, II, *Bollettino di bibliografia e storia delle scienze matematiche* 15, 47–53, 75–81.

163.1   (Unsigned) 'Mario Pieri', *A.p.I. Discussiones* 4, 31–5.

*164–18   'Quaestiones de grammatica', *A.p.I. Discussiones* 4, 41–4.

165–147   (With E. D'Ovidio) 'Relazione sulla memoria del Dr. Vincenzo Mago: Teoria degli ordini', *Atti Accad. sci. Torino* 49, 169.

l   *Academia pro Interlingua, Discussiones*, vol. 4 (Torino: Bocca).

1914

166–148   'Residuo in formulas de quadratura', *Mathesis* (4) 4, 5–10.

*167–19   *Fundamento de Esperanto* (Cavoretto (Torino)).

167.1   'Prof. Augusto Actis', *Revista universale* (U. Basso, ed.) 4, no. 39, 22.

167.2   'Prof. Louis Couturat', *Revista universale* (U. Basso, ed.) 4, no. 40, 78–9.

168–149   (With C. Segre) 'Relazione intorno alla memoria del Prof. Cesare Burali-Forti, Isomerie vettoriali e moti geometrici', *Atti Accad. sci. Torino* 50, 237.

1915

XIV, *169–20 *Vocabulario commune ad latino-italiano-français-english-deutsch*, 2nd ed. (Cavoretto (Torino): Academia pro Interlingua).

170–150 'Resto nella formula di Cavalieri-Simpson', *Atti Accad. sci. Torino* 50, 481–6.

170'–150' 'Residuo in formula de quadratura Cavalieri-Simpson', *Enseignement math.* 18 (1916), 124–9.

171–151 'Sul prodotto di grandezze', *Bollettino di matematiche e di scienze fisiche e naturali* 16, 99–100.

172–152 'Definitione de numero irrationale secundo Euclide', *Mathesis, società italiana di matematica, Bollettino* 7, 31–5.

*173–21 'Praepositiones internationale', *World-Speech* (Marietta, Ohio), no. 37, 2–3.

174–153 (With E. D'Ovidio) 'Relazione sulla memoria di G. Sannia, I limiti d'una funzione in un punto limite del suo campo', *Atti Accad. sci. Torino*, 50, 968–9.

175–154 'Le grandezze coesistenti di Cauchy', *Atti Accad. sci. Torino* 50, 1146–55.

176–155 'Importanza dei simboli in matematica', *Scientia* 18, 165–73.

176'–155' 'Importance des symboles en mathématique', *Scientia* 18, supplement, 105–14.

177–156 'Le definizioni per astrazione', *Mathesis, società italiana di matematica, Bollettino* 7, 106–20.

178–157 'L'esecuzione tipografica delle formule matematiche', *Atti Accad. sci. Torino* 51, 279–86. Reprinted in *Bollettino di matematica* 14, 121–8.

*179–22 'Bello et Lingua', *International Language*; a *Journal of 'Interlinguistics'* 5, 17–19.

1916

180–158 'Approssimazioni numeriche', *Atti Accad. naz. Lincei, Rend., Cl. sci. fis. mat. nat.* (5) 25–I, 8–14.

181–159 'Sul principio d'identità', *Mathesis, società italiana di matematica, Bollettino* 8, 40–1.

m 'Gli Stati Uniti della Terra', *Gazzetta del popolo della sera* (Torino), 8 March 1916.

1917

182–160 'Valori decimali abbreviati e arrontondati', *Atti Accad. sci. Torino* 52, 372–82. Reprinted in *Period. mat.* (3) 14, 97–105.

183–161 'Approssimazioni numeriche', *Atti Accad. sci. Torino* 52, 453–68, 513–28.

1918

184–162 'Eguale; Infinito; Logica matematica; Vettori', *Dizionario cognizioni utili* (and supplement) (Torino: Unione Tipografico-Editrice Torinese, 1917–19).

184'–162' 'Eguale', *Bollettino di matematica* 15 (1917–18), 195–8.
184"–162" 'Vettori', *Bollettino di matematica* 16 (1918–19), 157–8.
185–163 *Tavole numeriche* (Unione Tipografico-Editrice Torinese) (11th reprinting, 1963).
186–164 'Interpolazione nelle tavole numeriche', *Atti Accad. sci. Torino* 53, 693–716.
187–165 'Resto nelle formule di interpolazione', *Scritti offerti ad E. D'Ovidio* (Torino: Bocca), 333–5.
188–166 (With T. Boggio) 'Matteo Bottasso', *Mathesis, società italiana di matematica, Bollettino* 10, 87–8.

1919

189–167 'Sulla forma dei segni di algebra', *Giorn. mat. finanz.* 1, 44–9.
190–168 'Filiberto Castellano', *Mathesis, società italiana di matematica, Bollettino* 11, 62–3.
191–169 'Risoluzione graduale delle equazioni numeriche', *Atti Accad. sci. Torino* 54, 795–807.
192–170 'Conferenze matematiche', *Annuario dell'Università di Torino* (1919–20), 245–6.

1921

193–171 'Le definizioni in matematica', *Period. mat.* (4) 1, 175–89.
194–172 'Area de rectangulo', *Rassegna di matematica e fisica* (Rome, Istituto fisico-matematico G. Ferraris) 1, 200–3.
195–173 Review of T. Boggio, *Calcolo differenziale, con applicazioni geometriche. Esercitazioni matematiche*, (Circolo matematico, Catania) 1, 198–201.

1922

196–174 'Calculo super calendario', *A.p.I. Circulare* (1 January 1922), 2–3, 6–7.
196'–174' 'Calculo super calendario', *Urania; notizie di astronomia, meteorologia, chimica e fisica* 11, 20–4.
196"–174" 'Calculo super calendario', *Schola et vita* 3 (1928), 4–6.
197–173 'Operazioni sulle grandezze', *Atti Accad. sci. Torino* 57, 311–31.
197'–175' 'Operationes super magnitudines', *Rassegna di matematica e fisica* (Roma, Istituto fisico-matematico G. Ferraris) 2, 269–83.
*198–23 'Review of A. L. Guérard, *A Short History of the International Language Movement*', *A.p.I. Circulare* (1922), no. 2, 5–11.
*199–24 'Regulas pro Interlingua', *A.p.I. Circulare* (1922), no. 3, 1–3.
*200–25 'Lingua internationale ante Societate de Nationes', *A.p.I. Circulare* (1922), no. 4, 1–4.
201–176 'Theoria simplice de logarithmos', *Wiadom. mat.* 26, 53–5.
201.1 Letter to the Editor, *La matematica elementare* 1, 76.
201.2 'Problemi practici', *La matematica elementare* 1, 88–9.

1923

202–177 Review of A. Natucci, *Il concetto di numero, e le sue estensioni*, *Archeion, archivio di storia della scienza* 4, 382–3.

202.1 Review of Feder and Schneeberger, *Volständiges Wurzelwörterbuch Ido-Deutsch*, (Lüsslingen, 1919) and S. Auerbach, *Wörterbuch Deutsch-Ido* (Leipzig, 1922) *A.p.I. Circulare* (1923), no. 1, 6–8.

202.2 Review of *Mondo, Revuo por linguo Internaciona Ido*, *A.p.I. Circulare* (1923), no. 3, 7–8.

202.3 'Discussione', *A.p.I. Circulare* (1923), no. 4, 2–3.

1924

XV, 203–178 *Giochi di aritmetica e problemi interessanti* (Torino: Paravia) (Several later editions).

204–179 'I libri di testo per l'aritmetica nelle scuole elementari', *Period. mat.* (4) 4, 237–42.

*205–26 *Interlingua (Historia – Regulas pro Interlingua – De vocabulario – Orthographia – Lingua sine grammatica)* (Cavoretto (Torino)) (2nd ed., 1924; 3rd ed., 1927).

206–180 'De aequalitate', *Proceedings of the International Congress of Mathematicians, Toronto, 1924* (Toronto: University of Toronto Press, 1928), vol. 2, 988–9.

206'–180' 'De aequalitate', *A.p.I. Circulare* (1924), no. 5, 8–11.

XVI, *207–27 *Academia pro Interlingua, Circulares* (Torino, 1909–24).

1925

207.1 Letter to the Editor (A proposito dell'articolo di L. Siriati), *Bollettino di matematica* (2) 3, 129.

207.2 'International Auxiliary Language Association', *A.p.I. Circulare* (1925), no. 5, 76–7.

*208–28 'Pro historia de Interlingua, §1. Volapük', *A.p.I.* (1925), no. 5, 81–5.

n *Academia pro Interlingua* (Cavoretto (Torino)).

1926

*209–29 'Pro historia de Interlingua, §2. Academia in periodo 1893–1908', *A.p.I.* (1926), no. 2, 33–4.

*210–30 'Pro historia de Interlingua, §3. Latino simplificato', *A.p.I.* (1926), no. 4, 73–6.

*211–31 'Sinense', *A.p.I.* (1926), no. 6, 129–30.

212–181 'Quadrato magico', *Schola et vita* 1, 84–7.

213–182 'Jocos de arithmetica', *Schola et vita* 1, 166–73.

o *Academia pro Interlingua* (Cavoretto (Torino)).

1927

*214–32 Review of E. S. Pankhurst, *Delphos the Future of International Language*, *A.p.I.* (1927), no. 1, 14–20.

215–183 'De vocabulo matematica', *Rivista di matematica pura ed applicata per gli studenti delle scuole medie* 2, 212.

216–184 'Sulla riforma del calendario', *Atti Accad. sci. Torino* 62, 566–8.
217–185 'Historia et reforma de calendario', *A.p.I.* (1927), no. 3, 49–55.
218–186 'Vocabulario matematico', *Rivista di matematica pura ed applicata per gli studenti delle scuole medie* 3, 270–2.
    p *Academia pro Interlingua* (Cavoretto (Torino)).

1928

219–187 'Gio. Francesco Peverone ed altri matematici piemontesi ai tempi di Emanuele Filiberto', *Studi pubblicati dalla R. Università di Torino nel IV centenario della nascita di Emanuele Filiberto* (Torino: R. Università di Torino).
220–188 'Historia de numeros', *Schola et vita* 3, 139–42. Reprinted in *Archeion*; *archivio di storia della scienza* 9, 364–6.
  220.1 'Ad omne interlinguista collega et amico, gratias', *Schola et vita* 3, 201.
221–189 'Interessante libro super calculo numerico', *Schola et vita* 3, 217–9.
222–190 (With Cesarina Boccalatte) 'La geometria basata sulle idee di punto e angolo retto', *Atti Accad. sci. Torino, Cl. sci. fis. mat. nat.* 64–I, 47–55.

1929

*223–33 'Vocabulos internationale', *Schola et vita* 4, 58–61.
*224–34 'Volapük post 50 anno', *Schola et vita* 4, 225–33.
  224.1 Letter to the Editor, *Time and Tide* (London, 5 July 1929). Reprinted in *Schola et vita* 4, 247–8.
225–191 'Monete italiane nel 1929', *Giornale di matematica e fisica* 3, no. 4.

1930

*226–35 'Studio de linguas', *Rendiconti della Unione professori*, Milano (April 1930), 1–2. Reprinted in *Schola et vita* 5, 81–4.
*227–36 'Quaestiones de interlingua: ablativo aut nominativo', *Schola et vita* 5, 138–40.
228–192 (With Fausta Audisio) 'Calcolo di π colla serie di Leibniz', *Atti Accad. naz. Lincei, Rend., Cl. sci. fis. mat. nat.* (6) 11–I, 1077–80.
*229–37 'Algebra de grammatica', *Schola et vita* 5, 323–36.

1931

230–193 'Jocos de arithmetica', *Rendiconti della Unione professori*, Milano (October 1931), 50–1.
*231–38 'Libertate et Unione', *Schola et vita* 6, 323–5.

BIBLIOGRAPHY

Agostinelli, Cataldo: 1965, 'Tommaso Boggio', *Atti Accad. sci. Torino, Cl. sci. fis. mat. nat.* 99, 281–296.

Aimonetto, Italo: 1969, 'Il concetto di numero naturale in Frege, Dedekind e Peano', *Filosofia* 20, 579–606.

Anon.: 1922, *Della Vita e degli Studi di Conte Carlo Ceppi, Architetto Torinese*, Torino (E. Celanze).

Ascoli, Guido: 1932, (untitled necrology of G. Peano) *La ricerca scientifica* 3, 592–3.

Ascoli, Guido: 1955, 'I motivi fondamentale dell'opera di Giuseppe Peano' [Terracini 1955], 23–30.

Barone, Francesco: 1955, 'Un'apertura filosofica della logica simbolica peaniana' [Terracini 1955], 41–50.

Berteu, Agostino: 1898, *Vita dell'Abate Francesco Faà di Bruno*, Torino.

Bertoli, Paulus: 1971, 'Beatificationis et canonizationis Servi Dei Francisci Faà di Bruno, . . . ', *Acta Apostolicae Sedis* 63, 788–790.

Birkhoff, Garrett, ed.: 1973, *A source book in classical analysis*, Cambridge, Mass. (Harvard University Press).

Boggio, Tommaso: 1933, 'Giuseppe Peano', *Atti Accad. sci. Torino, Cl. sci. fis. mat. nat.* 68, 436–46.

Boggio, Tommaso: 1955, 'Il calcolo geometrico di Peano' [Terracini 1955], 65–9.

Botto, Costantino: 1934, 'Un'autentica gloria cuneese e italiano: il matematico Giuseppe Peano, Cuneo 1858 – Torino 1932', *Annuario del Istituto Tecnico di Cuneo* (1933–34), 5–24.

Boulton, Marjorie: 1960, *Zamenhof, Creator of Esperanto*, Atlantic Highlands, N. J. (Humanities Press).

Bruno, Giuseppe: 1886, 'Donato Levi', *Annuario dell'Università di Torino* (1885–86), 132–133.

Camerano, Lorenzo: 1895, 'Michele Lessona', *Annuario dell'Università di Torino* (1894–95), 181–191.

Carruccio, Ettore: 1955, 'Spunti di storia delle matematiche e della logica nell'opera di G. Peano' [Terracini 1955], 103–14.

Cassina, Ugo: 1928, 'In occasione de septuagesimo anno de Giuseppe Peano', *Schola et Vita* (supplement 27 August 1928), 5–28.

Cassina, Ugo: 1932, 'Vita et opera de Giuseppe Peano', *Schola et Vita*, 7, 117–48.

Cassina, Ugo: 1933a, 'Su la logica matematica di G. Peano', *Boll. Un. mat. ital.* 12, 57–65 (Also in [Cassina 1961a]).

Cassina, Ugo: 1933b, 'L'opera scientifica di Giuseppe Peano', *Rendiconti del Seminario matematico e fisico di Milano* 7, 323–89 (Also in [Cassina 1961b]).

Cassina, Ugo: 1933c, 'L'oeuvre philosophique de G. Peano', *Revue de métaphysique et de morale* 40, 481–91.

Cassina, Ugo: 1937, 'Parallelo fra la logica teoretica di Hilbert e quella di Peano', *Period.*

211

mat. (4) 17, 129–38 (Also in [Cassina 1961a]).

Cassina, Ugo: 1950a, 'Il concetto di linea piana e la curva di Peano', *Rivista di matematica della Università di Parma* 1, 275–92.

Cassina, Ugo: 1950b, 'L'area di una superficie curva nel carteggio inedito di Genocchi con Schwarz ed Hermite', *Istituto Lombardo, Accademia di scienze e lettere, Rendiconti, Scienze* 83, 311–28 (Also in [Cassina 1961b]).

Cassina, Ugo: 1952, 'Alcune lettere e documenti inediti sul trattato di Calcolo di Genocchi-Peano', *Istituto Lombardo, Accademia di scienze e lettere, Rendiconti, Scienze* 85, 337–62 (Also in [Cassina 1961b]).

Cassina, Ugo: 1953a, 'Giovanni Vacca', *Istituto Lombardo, Accademia di scienze e lettere, Parte gen. e Atti uff.* 86, 185–200.

Cassina, Ugo: 1953b, 'Sulla critica di Grandjot all'aritmetica di Peano', *Boll. Un. mat. Ital.* (3) 8, 442–7 (Also in [Cassina 1961a]).

Cassina, Ugo: 1953c, 'L'idéographie de Peano du point de vue de la théorie du language', *Rivista di matematica della Università di Parma* 4, 195–205 (Italian translation in [Cassina 1961a]).

Cassina, Ugo: 1953d, 'Su l'opera filosofica e didattica di Giuseppe Peano', [Cassina 1961a], 343–57 (Talk given in Cuneo, 6 December 1953).

Cassina, Ugo: 1955a, 'Storia ed analisi del 'Formulario completo' di Peano', *Boll. Un. mat. Ital.* (3) 10, 244–65, 544–74 (Also in [Cassina 1961b]).

Cassina, Ugo: 1955b, 'Sul *Formulario mathematico* di Peano', [Terracini 1955], 71–102 (Also in [Cassina 1961a]).

Cassina, Ugo: 1957, 'Un chiarimento sulla biografia di G. Peano', *Boll. Un. mat. Ital.* (3) 12, 310–12.

Cassina, Ugo: 1958, 'Su un teorema di Peano e il moto del polo', *Istituto Lombardo, Accademia di scienze e lettere, Rendiconti, Scienze* 92, 631–55.

Cassina, Ugo: 1961a, *Critica dei principî della matematica e questioni di logica*, Rome (Cremonese).

Cassina, Ugo: 1961b, *Dalla geometria egiziana alla matematica moderna*, Rome (Cremonese).

Cicognani, C.: 1955, Beatificationis et canonizationis Servi Dei Francisci Faà di Bruno, . . . , *Acta Apostolicae Sedis* 47, 750–2.

Clark, Ronald W.: 1976, *The Life of Bertrand Russell*, New York (Knopf).

Comanducci, A. M.: 1934, *I Pittori Italiani dell'Ottocento. Dizionario Critico e Documentario*, Milan (Casa Editrice Artisti D'Italia).

Couturat, Louis: 1899, 'La logique mathématique de M. Peano', *Revue de métaphysique et de morale* 7, 616–46.

Couturat, Louis: 1901, (Analytical review of Peano, *Formulaire de mathématique*. Introduction and volumes I–III), *Bulletin des sciences mathématiques* (2) 25–I, 141–59.

Della Casa, Luciano: 1933, 'Infantia et juventute de Giuseppe Peano', *Schola et Vita* 8, 141–4.

Dictionary of Scientific Biography, C. C. Gillispie, ed., 14 volumes (1970–76), New York (Charles Scribner's Sons).

Dickstein, Samuel: 1934, 'Peano jako historyk', *Wiadom. mat.* 36, 65–70.

D'Ovidio, Enrico: 1889, 'Francesco Faà di Bruno', *Annuario dell'Università di Torino* (1888–89), 156–64.

D'Ovidio, Enrico: 1892, 'Angelo Genocchi', *Atti Accad. sci. Torino, Cl. sci. fis. mat. nat.* 27, 1090–106.
Dudman, Victor: 1971, 'Peano's Review of Frege's *Grundgesetze*', *Southern Journal of Philosophy* 9, 25–37.
Enriques, Federigo: 1922, *Per la storia della logica*, Bologna.
Fano, Gino: 1933, 'Enrico D'Ovidio', *Annuario dell'Università di Torino* (1932–33), 443–9.
Ferraris, Galileo: 1895, 'Giuseppe Basso', *Atti Accad. sci. Torino, Cl. sci. fis, mat. nat.* 31, 3–17.
Feys, Robert: 1953, 'Peano et Burali-Forti précurseurs de la logique combinatoire', *Actes du XIème Congrès international de philosophie*, vol. 5 (Amsterdam, Louvain), 70–2.
Frege, Gottlob: 1896, 'Ueber die Begriffsschrift des Herrn Peano und meine eigene', *Berichte über die Verhandlungen der Königlich Sächsischen Gesellschaft der Wissenschaften zu Leipzig, mathematisch-physikalische Klasse* 48, 361–78.
Frege, Gottlob: 1976, *Nachgelassene Schriften und Wissenschaftlicher Briefwechsel*, Hamburg (Felix Meiner Verlag).
Gabba, Alberto: 1957, 'La definizione di area di una superficie curva ed un carteggio inedito di Casorati con Schwarz e Peano', *Istituto Lombardo, Accademia di scienze e lettere, Rendiconti, Scienze* 91, 857–83.
Geymonat. Ludovico: 1955, 'I fondamenti dell'aritmetica secondo Peano e le obiezioni 'filosofiche' di B. Russell' [Terracini 1955], 51–63.
Geymonat, Ludovico: 1959, 'Peano e le sorti della logica in Italia', *Boll. Un. mat. Ital.* (3) 14, 109–18.
Geymonat, Ludovico: 1971–72, *Storia del pensiero filosofico scientifico*, Milan (Garzanti).
Gliozzi, Mario: 1932, 'Giuseppe Peano', *Archeion; archivio di storia della scienza* 14, 254–5.
Grattan-Guinness, I.: 1975, 'Wiener on the Logics of Russell and Schröder. An Account of his Doctoral Thesis and of his Discussion of it with Russell', *Annals of Science* 32, 103–82.
Grattan-Guiness, I.: 1977, *Dear Russell – Dear Jourdain: A commentary on Russell's logic, based on his correspondence with Philip Jourdain*, New York (Columbia University Press).
Guareschi, Icilio: 1917, 'Ugo Schiff', *Atti Accad. sci. Torino, Cl. sci. fis. mat. nat.* 52, 333–52.
Heijenoort, Jean van, ed.: 1967, *From Frege to Gödel: A Source Book in Mathematical Logic, 1879–1931*, Cambridge, Mass. (Harvard University Press).
Jourdain, P. E. B.: 1912, 'The Development of the Theories of Mathematical Logic and the Principles of Mathematics', *Quarterly Journal of Pure and Applied Mathematics* 43, 219–314 (pp. 270–314 treat of Peano's work).
Kennedy, H. C.: 1963, 'The mathematical philosophy of Giuseppe Peano', *Philosophy of Science* 30, 262–6.
Kennedy, H. C.: 1968, 'Giuseppe Peano at the University of Turin', *The Mathematics Teacher* 61, 703–6.
Kennedy, H. C.: 1970, 'Cesare Burali-Forti', *Dictionary of Scientific Biography* 2, 593–4.

Kennedy, H. C.: 1972, 'The origins of modern axiomatics: Pasch to Peano', *The American Mathematical Monthly* 79, 133–6.

Kennedy, H. C.: 1973, 'What Russell learned from Peano', *Notre Dame Journal of Formal Logic* 14, 367–72.

Kennedy, H. C.: 1974a, *Giuseppe Peano*, Basel (Birkhäuser Verlag) (German translation by Ruth Amsler).

Kennedy, H. C.: 1974b, 'Alessandro Padoa', *Dictionary of Scientific Biography* 10, 274.

Kennedy, H. C.: 1974c, 'Giuseppe Peano', *Dictionary of Scientific Biography* 10, 441–4.

Kennedy, H. C.: 1974d, 'Mario Pieri', *Dictionary of Scientific Biography* 10, 605–6.

Kennedy, H. C.: 1974e, 'Peano's concept of number', *Historia Mathematica*1, 387–408.

Kennedy, H. C.: 1975, 'Nine letters from Giuseppe Peano to Bertrand Russell', *Journal of the History of Philosophy* 13, 205–20.

Kennedy, H. C.: 1976, 'Giovanni Vailati', *Dictionary of Scientific Biography* 13, 550–1.

Kozłowski, Wł.M.: 1934, 'Wspomnienie o Jósefa Peano', *Wiadom. mat.* 36, 57–64.

Levi, Beppo.: 1932, 'L'opera matematica di Giuseppe Peano', *Boll. Un. mat. Ital.* 11, 253–62.

Levi, Beppo: 1933, 'Intorno alle vedute di G. Peano circa la logica matematica', *Boll. Un. mat. Ital.* 12, 65–8.

Levi, Beppo.: 1955, 'L'opera matematica di Giuseppe Peano' [Terracini 1955], 9–21.

Mangione, Corrado: 1971, 'Logica e problema dei fondamenti nelle seconda metà dell'Ottocento' [Geymonat 1971] 5, 755–830.

Mangione, Corrado: 1972, 'La logica nel ventesimo secolo' [Geymonat 1972] 6, 469–682.

Medvedev, F. A.: 1965, 'Funktsii mnozhestva u [G.] Peano', *Istoriko-Matematicheskie Issledovaniya* 16, 311–23.

Medvedev, F. A.: 1974, *Razvitie ponyatiya integrala*, Moscow (Nauka).

Medvedev, F. A.: 1975, *Ocherki istorii teorii funktsii deistvitel'nogo peremennogo*, Moscow (Nauka).

Monna, A. F.: 1973, *Functional analysis in historical perspective*, New York (Wiley).

Morera, Giacinto: 1908, 'Francesco Siacci', *Atti Accad. sci. Torino, Cl. sci. fis. mat. nat.* 43, 568–78.

Naccari, Andrea: 1896, 'Giuseppe Basso', *Annuario dell'Università di Torino* (1895–96), 153–61.

Natucci, A.: 1932, 'In memoria di G. Peano', *Bollettino di matematica* (2) 11, 52–6.

Nidditch, Peter: 1963, 'Peano and the Recognition of Frege', *Mind* 72, 103–10.

Padoa, A.: 1933, 'Il contributo di G. Peano all'ideografia logica', *Period. mat.* (4) 13, 15–22.

Padoa, A.: 1936, 'Ce que la logique doit à Peano', *Actualités scientifiques et industrielles* 395, 31–7.

Parona, C. F.: 1912, 'Giorgio Spezia', *Annuario dell'Università di Torino* (1911–12), 281–42.

Russell, Bertrand: 1959, *My Philosophical Development*, London (Unwin).

Russell, Bertrand: 1967, *The Autobiography of Bertrand Russell, 1872–1914*, Boston (Little, Brown).

Segre, Beniamino: 1955, 'Peano ed il Bourbakismo' [Terracini 1955], 31–9.

Segre, Corrado: 1894, 'Giuseppe Bruno', *Annuario dell'Università di Torino* (1893–94), 155–68.

Stamm, E.: 1934, 'Jósef Peano', *Wiadom. mat.* **36**, 1–56.

Terracini, Alessandro, ed.: 1955, *In Memoria di Giuseppe Peano*, Cuneo (presso il liceo scientifico statale).

Terracini, Alessandro: 1968, *Ricordi di un matematico*, Rome (Cremonese).

Tricomi, F. G.: 1962, 'Matematici italiani del primo secolo dello stato unitario', *Memorie dell'Accad. sci. Torino*, Ser. 4, n. 1.

Tricomi, F. G.: 1967a, *La mia vita di matematico*, Padova (CEDAM).

Tricomi, F. G.: 1967b, 'Matematici torinesi dell'ultimo secolo', *Atti Accad. sci. Torino, Cl. sci. fis. mat. nat.* **102**, 253–278.

Tricomi, F. G.: 1969, 'Uno sguardo allo sviluppo della matematica in Italia nel primo secolo dello stato unitario', *Rendiconti del Seminario Matematico dell'Università e Politecnico di Torino* **28**, 63–76.

Tricomi, F. G.: 1972, 'Ricordi di mezzo secolo di vita matematica torinese', *Rendiconti del Seminario Matematico dell'Università e Politecnico di Torino* **31**, 31–43.

Trinchero, Mario: 1964, 'La fortuna di Frege nell'Ottocento', *Rivista di filosofia* **55**, 154–86.

Tripodi, Angelo: 1970, 'Considerazioni sull'epistolario Frege-Peano', *Boll. Un. mat. Ital.* (4) **3**, 690–8.

Vacca, Giovanni: 1932, 'Lo studio dei classici negli scritti matematici di Giuseppe Peano', *Atti della Società italiana per il progresso delle scienze* **21**, II, 97–9.

Vailati, Giovanni: 1899, 'La logique mathématique et sa nouvelle phase de développement dans les écrits de M. J. Peano', *Revue de métaphysique et de morale* **7**, 86–102.

Vivanti, Giulio: 1932, 'Giuseppe Peano', *Istituto Lombardo di scienze e lettere, Rendiconti* (2) **65**, 497–8.

Volterra, Vito: 1897, 'Giuseppe Erba', *Annuario dell'Università di Torino 1896–97*, 145–8.

Wiener, Norbert: 1953, *Ex-Prodigy*, New York (Simon and Schuster).

INDEX OF THE PUBLICATIONS OF GIUSEPPE PEANO

Numbers in brackets refer to the Chronological List of the Publications of Giuseppe Peano.

[1] 7–8	[37] 37, 44	[68] 41, 52
[2] 7–8	[38] 38	[69] 54
[3] 7–8	[38.1] 38	[70] 54
[4] 7–8	[39] 37	[70.1] 54
[5] 11, 15	[40] 39	[70.2] 54
[6] 11	[41] 39	[70.3] 46
[7] 15	[42] 39	[71] 45, 47
[8] 11–15	[42.1] 40	[72] 47
[9] 17	[43] 40	[73] 47
[10] 17–18, 20, 78	[44] 40	[74] 47
[11] 18–19	[44.1] 40	[75] 47
[12] 18, 20–21	[45] 41	[76] 54
[13] 21	[46] 40	[77] 56
[14] 21–24	[46.1] 40	[78] 55
[14.1] 24	[47] 40	[79] 57, 61
[15] 25	[48] 40	[80] 58, 61
[16] 25–27, 67, 108	[49] 41	[81] 64
[17] 28, 54, 69	[50] 41	[82] 74
[18] 27, 51, 73	[51] 40	[83] 64
[19] 28, 39	[52] 40, 64	[84] 59
[20] 28	[53] 40	[85] 65
[21] 12, 28	[54] 39	[86] 65
[22] 31	[55] 42	[86.1] 65
[23] 10	[56] 42	[87] 61
[24] 31–32	[57] 43	[88] 65
[25] 32	[58] 42–43	[89] 60
[26] 32	[58.1] 43	[90] 66
[27] 33	[59] 43	[90.1] 67
[28] 35	[60] 41, 43, 100	[90.2] 67
[29] 35	[61] 43	[90.3] 67
[30] 35	[62] 51	[90.4] 67
[31] 36, 73	[63] 51	[90.5] 67
[32] 36	[64] 51	[91] 68
[33] 38	[64.1] 52	[92] 69
[34] 38	[65] 52	[93] 64, 71, 76
[35] 36	[66] 45–46	[94] 78
[36] 38	[67] 52	[95] 78

[96] 78
[97] 76, 78
[98] 79
[99] 80
[100] 80
[101] 80
[102] 71
[103] 81
[103.1] 82
[103.2] 82
[103.3] 83
[103.4] 83
[104] 82
[105] 82
[106] 83, 108
[106.1] 82
[107] 77, 92
[108] 92
[109] 93
[110] 100
[111] 102
[112] 102
[113] 103
[114] 102
[115] 102
[116] 103, 105
[116.1] 103
[117] 103
[118] 81, 104
[119] 104
[120] 102
[121] 104
[122] 105
[123] 102
[124] 102
[125] 105
[126] 107
[127] 107
[127.1] 109
[128] 110
[129] 110
[130] 110
[131] 102
[131.1] 112
[132] 113
[133] 33, 118–120
[134] 120

[135] 120
[136] 120
[137] 111, 121
[137.1] 121
[137.2] 121
[138] 111, 121
[139] 127
[139.1] 129
[140] 124
[141] 110, 127, 129
[142] 130
[142.1] 130
[143] 131
[144] 131
[145] 133
[146] 129, 133
[147] 133
[148] 133
[149] 136
[150] 136
[151] 136
[152] 136
[153] 136
[154] 136
[154.1] 137
[155] 137
[156] 137
[157] 138
[158] 137
[159] 138–139
[159.1] 139
[159.2] 139
[159.3] 139
[159.4] 139
[160] 140
[161] 140
[162] 140
[163] 136, 141
[163.1] 142
[164] 142
[165] 142
[166] 142
[167] 143
[167.1] 142
[167.2] 144
[168] 144
[169] 110, 130, 144

[170] 145
[171] 145
[172] 146
[173] 147
[174] 146
[175] 19, 146
[176] 148
[177] 148
[178] 148
[179] 147
[180] 148
[181] 149
[182] 149
[183] 149
[184] 149
[185] 150
[186] 150
[187] 151
[188] 151
[189] 152
[190] 34, 85, 151
[191] 152
[192] 145, 153
[193] 93, 154
[194] 154
[195] 154
[196] 155
[197] 155
[198] 155
[199] 155
[200] 155
[201] 156
[201.1] 156
[201.2] 156
[202] 157
[202.1] 157
[202.2] 157
[202.3] 157
[203] 158–159
[204] 159
[205] 160
[206] 160
[207] 133
[207.1] 161
[207.2] 162
[208] 162
[209] 162

[210] 162
[211] 162
[212] 163
[213] 163
[214] 163
[215] 163
[216] 164
[217] 164

[218] 163
[219] 164
[220] 164
[221] 165
[222] 165
[223] 166
[224] 128, 166
[224.1] 166

[225] 166
[226] 167
[227] 167
[228] 167
[229] 167
[230] 167
[231] 167

INDEX OF NAMES

Abel, Niels Henrik (1802–1829) 109
Actis, Augusto (1831–1914) 142
Adler, Felix (1851–1933) 92
Agnesi, Maria Gaetana (1718–1799) 65
Alloati, G. B. 171
Amaldi, Ugo (1875–1957) 165
Amanzio, Domenico (1854–1908) 40
Amodeo, Federico (1859–1946) 187
Anselmi, Giorgio 17, 20
Andreoli, Giulio 191
Appell, Paul (1855–1930) 57
Aquinas, Thomas (ca. 1225–1274) 149
Arbicone, A. 103
Archimedes (ca. 287 B.C.–212 B.C.) 101–102
Aristotle (384 B.C.–322 B.C.) 49, 71, 104, 149, 154
Arrighi, G 193
Arzelà, Cesare (1847–1912) 29, 104
Ascoli, Guido (1887–1957) 170
Astuti, Teresa 190
Audisio, Fausta 167, 187, 193

Babcock, Earle B. (1881–1935) 162
Bach, Johann Sebastian (1685–1750) 85
Bagni 102
Baldwin, James Mark (1861–1934) 92, 96
Barale, Michele 3
Barbarin, Paul (1855–1931) 110
Barberis, Antonio 6
Bardelli, Giuseppe (1837–1908) 42
Basso, Giuseppe (1842–1895) 4, 6, 182–184
Basso, Ugo 142
Battaglini, Giuseppe (1826–1894) 176
Beethoven, Ludwig van (1770–1827) 85
Bellavitis, Giusto (1803–1880) 22, 181
Beltrami, Eugenio (1835–1899) 35, 59, 176

Benedict XIII (Pedro de Luna) (1394–1417) 109
Bernardi, Luigi 1
Bernhaupt, J. 142
Bernoulli, Daniel (1700–1782) 72
Bernoulli, Jakob (1654–1705) 72
Bernoulli, Johann (1667–1748) 72
Bernstein, Felix (1878–1956) 118
Bersano, Carlo 187, 192
Berteu, Agostino 79
Berzolari, Luigi (1863–1949) 166
Bettazzi, Rodolfo (1861–1941) 78, 80, 83, 101, 103, 187–188
Bijlevelt, Bonto van 128
Blaserna, Pietro (1836–1918) 31, 112
Boccalatte, Cesarina 165, 187, 193
Boccardi, Giovanni (1859–1936) 98
Boggio, Tommaso (1877–1963) 84, 88, 101, 103, 109, 124, 134, 145, 151, 154, 156, 160, 170, 187, 189–192
Bohlmann, Georg (1869–1928) 15
Bolyai, János (1802–1860) 176
Boncompagni, Baldassare (1821–1894) 186
Bon-Compagni, Count Carlo 5
Bongiovanni, Emilia 164
Bonnet, Ossian (1819–1892) 16
Boole, George (1815–1864) 21, 27–28, 46, 73, 76
Borchardt, Carl Wilhelm (1817–1880) 69
Borel, Emile (1871–1956) 118, 122, 131
Borio, Agostino 156, 187
Bosco, Giovanni (1815–1888) 185
Bottasso, Matteo (1878–1918) 145, 151, 187, 190–192
Botto, Costantino 100–101
Bouquet, Jean-Claude (1819–1885) 17
Bourbaki, Nicolas 174

Boussinesq, Joseph Valetin (1842–1929) 30–31, 35, 93

Bouton, Charles L. (1869–1922) 70

Boutroux, Pierre Léon (1880–1922) 111

Boverio, Ernesto 191

Briggs, Le Baron Russell (1855–1934) 134

Brioschi, Francesco (1824–1897) 59–60

Briot, Charles Auguste (1817–1882) 17

Bruno, Giordano (1548–1600) 25

Bruno, Giuseppe (1828–1893) 3, 5, 11, 183–184

Burali-Forti, Cesare (1861–1931) 23, 45, 47, 49, 53, 71, 78, 80, 82–86, 88–89, 91, 94, 97, 118, 134, 144, 155, 160, 187–192

Calderoni, Mario (1879–1914) 96, 112, 130

Canesi, Gaetano 166, 169, 170

Cantelli, Francesco P. (1875–1966) 191

Cantoni, Carlo 96, 112

Cantoni, E. 102–103

Cantor, Georg (1845–1918) 18, 31, 40, 61–63, 69–70, 86, 92, 102, 118–119, 122, 134

Cantor, Moritz (1829–1920) 47, 92

Carus, Paul (1852–1919) 92, 96, 123

Carvallo, Emmanuel M. (1856–1945) 39, 43

Casorati, Felice (1835–1890) 12–13, 29–31

Cassina, Ugo (1897–1964) 14, 32, 61, 83–84, 87, 111, 121, 145, 155–156, 161, 163, 165–167, 170–174, 187, 192–193

Cassinis, Gino (1885–1964) 167

Castellano, Filiberto (1860–1919) 34, 36, 42, 44–45, 47, 49, 54, 80, 83–85, 90, 103, 139, 151, 187–189, 191

Catalan, Eugène Charles (1814–1894) 52

Catania, Sebastiano 121, 191

Cauchy, Augustin-Louis (1789–1857) 14, 19, 24, 78–79, 131, 181, 185

Cavalieri, Bonaventura (1598–1647) 101, 145

Cavallo, G. Michele 1

Cayley, Arthur (1821–1895) 46, 176

Ceppi, Carlo Giulio (1829–1921) 3–4, 177–178

Certo, Luigi 80

Cesi, Federico (1585–1630) 112

Chanaud, Jean 171

Chandler, Seth Carlo (1846–1913) 57–58

Chevalier, Maurice (1888–1971) 168

Chinaglia, Piera 164, 187

Chini, Mineo (1866–1933) 83, 104, 187–189

Chiò, Felice (1813–1871) 181, 183

Chionio, Florentio 187

Ciamberlini, Corrado (1861–1944) 102, 187

Cibrario Cinquini, Maria 165, 187, 193

Cipolla, Michele (1880–1947) 187

Cisotti, Umberto (1882–1946) 167

Ciuropajlowicz, T. 162

Clifford, W. K. (1845–1879) 46

Colbert, Claudette (1907–) 168

Comi, Tiziana Tersilla 187, 192

Conti, Alberto (1873–1940) 103

Couturat, Louis (1868–1914) 91, 94–96, 98, 100, 108, 111–112, 120, 123, 127–129, 136, 144, 157

Creighton, James Edwin (1861–1924) 92

Cremona, Luigi (1830–1903) 70

Crispi, Francesco (1819–1901) 66

Crosio, Aneta 20

Crosio, Bianca 20

Crosio, Camilla 20

Crosio, Carola (?–1940) 20, 42

Crosio, Luigi (1835–1915) 20

D'Annunzio, Gabriele (1863–1938) 153

Darboux, Jean-Gaston (1842–1917) 98

D'Arcais, Francesco (1849–1927) 29, 38, 65

Darwin, Charles (1809–1882) 179

Darwin, George Howard (1845–1912) 58

De Amicis, Edmondo (1846–1908) 176
De Bosis, Virginia 88
Dedekind, Richard (1831–1916) 25–26, 37, 74, 82, 92, 118, 140, 174
De Filippi, Filippo (1814–1867) 179
Delboeuf, J. 46
Del Grosso, Remigio (1813–1876) 176
Della Casa, Luciano 187, 191
Dell'Agnola, Carlo Alberto (1871–1956) 190
De Morgan, Augustus (1806–1871) 46
Descartes, René (1596–1650) 49
Destefanis, Maria 187
Dickstein, Samuel (1851–1939) 38, 66, 96, 156, 165, 169
Dini, Ulisse (1845–1918) 24–25, 102
Dirichlet, Gustav Peter Lejeune (1805–1859) 15, 102
Dorna, Alessandro (1825–1887) 36
D'Ovidio, Enrico (1842–1933) 3–8, 11, 16, 42, 65, 78, 80, 104, 110, 124, 132, 142, 146, 149, 151, 154, 156, 166, 170, 176–177, 182
D'Ovidio, Francesco (1849–1925) 177
Duhem, Pierre (1861–1916) 122
Dumas 69
Duprez, Marcel 57
Duse, Eleonora (1859–1924) 42
Dusi, Teresa 153

Einstein, Leopold 128
Ellis, A. J. (1814–1890) 46
Emanuele Filiberto (1528–1580) 164
Eneström, Gustav Hjalmar (1852–1923) 58, 102
Enriques, Federigo (1871–1946) 152, 166, 172
Erba, Giuseppe Bartolomeo (1819–1895) 5, 184–185
Euclid (fl. ca. 295 B.C.) 26, 49, 52, 82, 102, 132, 146, 154–155
Euler, Leonhard (1707–1783) 58, 72, 101, 181

Faà di Bruno, Francesco (1825–1888) 5, 8, 79, 185
Fano, Gino (1871–1952) 41, 47, 83, 103, 187
Fantappiè, Luigi (1901–1956) 167
Favre, Emilio 3
Fazzari, Gaetano (1856–1935) 103
Fergola, Emanuele (1831–1915) 176
Fermat, Pierre de (1601–1665) 103, 162
Farrari, Ettore (1845–1929) 25
Ferrati, Camillo (1822–1888) 5, 184
Ferrero, Clementina 161, 187, 192
Forel, Auguste (1848–1931) 147
Foresti, Luigi 3
Foster, Edward P. 148
Francis Ferdinand (1863–1914) 144
Franel, Ami Jérôme (1859–1939) 69, 71
Frege, Gottlob (1848–1925) 15, 26, 36, 46, 64, 69, 73–78, 92, 118, 134, 174
Freudenthal, Hans (1905–) 91, 95
Frisone, Rosetta 187, 191–192
Fubini, Guido (1879–1943) 124, 132, 160, 170

Gabba, Alberto 166
Galilei, Galileo (1564–1642) 101, 112
Garibaldi, Giuseppe (1807–1882) 9
Gauss, Carl Friedrich (1777–1855) 102, 181
Geiser, Carl Friedrich (1843–1934) 69–70, 72
Genocchi, Angelo (1817–1889) 4, 8–14, 16–18, 20, 28, 30, 177, 180–182
Gentile, Giovanni (1875–1944) 156–157
Gérard, L. 104
Gerbaldi, Francesco (1858–1934) 7, 62, 177
Gergonne, Joseph Diaz (1771–1859) 109
Gessel 58
Geymonat, Ludovico (1908–) 93
Giacosa, Giuseppe (1847–1906) 176
Gilbert, Philip 15–16, 41
Giraldo, Lucia 1
Giudice, Francesco (1855–1936) 24–25, 39–41, 47, 65, 78, 80, 83, 187–191

Gliozzi, Mario 193
Gmeiner, Josef Anton (1862–1927) 103
Govi, Gilberto (1826–1889) 183
Gramegna, Maria (1887–1915) 132, 187, 190
Grassmann, Hermann (1809–1877) 18, 21–23, 38–39, 43, 46, 54, 60, 64
Guérard, Albert Leon 155
Guglielmi, Amalia 156
Günther, Siegmund 46
Guyou, Emile (1843–1915) 56
Gyldén, Hugo (1841–1896) 58

Halsted, G. B. (1853–1922) 46, 97
Hamilton, William Rowan (1805–1865) 22
Harkness, James (1864–1923) 70
Hausdorff, Felix (1868–1942) 32
Heijenoort, Jean van (1912–) 73
Hermite, Charles (1822–1901) 10, 14, / 31, 69, 185
Hertz, H. R. (1857–1897) 60
Hilbert, David (1862–1943) 27, 32, 86, 97, 110
Hill, George William (1838–1914) 70
Hjelslev, Johannes T. (1873–1950) 166
Hoepli, Ulrico (1847–1935) 71
Holmes, M. A. F. 127, 129
Horace (Quintus Horatius Flaccus) (65 B.C.–8 B.C.) 119
Houël, Guillaume-Jules (1823–1886) 15
Hurwitz, Adolf (1859–1919) 69–71

Itelson, G. 138

Jadanza, Nicodemo (1847–1920) 36, 43, 149, 184
James, William (1842–1910) 84
Jespersen, Otto (1860–1943) 166
Jevons, W. S. (1835–1882) 46
Jordan, Camille (1838–1922) 15–16, 18, 67
Jourdain, P. E. B. (1879–1919) 26, 132–134

Kepler, Johannes (1571–1630) 149
Kerckhoffs, Auguste 125–126

Kirchhoff 126
Klein, Felix (1849–1925) 20, 53, 69–72, 92
Koch, Helge von (1870–1924) 32
Korzybski, Alfred Habdank Skarbek (1879–1950) 160
Kovalevskaya, Sotya [Ковалевская, С.В.] (1850–1891) 40
Kryzhanovskii, D.A. [Крыжановский, Д.А.] (1883–1938) 170
Küstner, F. 58

Ladd Franklin, Christine (1847–1930) 96, 123
Lagneau 192
Lagrange, Joseph-Louis (Giuseppe Luigi) (1736–1813) 101, 181
Laisant, Charles Ange (1841–1920) 70, 98
Lampe, Emil (1840–1908) 188
Lanner, Alois 66
Lantelme, Giuseppe 5, 184
Laura, Ernesto (1879–1949) 104, 187
Laurent, Hermann (1841–1908) 40, 42
Lazzarino, Orazio 192
Lazzeri, Giulio (1861–1935) 103
Leau, Leopold 96, 98, 112
Lebesque, Henri (1875–1941) 18–19, 174
Lecornu, Léon (1854–1940) 54
Leibniz, Gottfried Wilhelm (1646–1716) 36, 46–47, 65, 67, 71, 83, 87, 91, 107–108, 125, 148
Lemoine, Emile (1840–1912) 70, 110
Leonardo da Vinci (1452–1519) 49
Leonardo of Pisa (Fibonacci) (ca. 1170– after 1240) 156
Lerch, Mathias (1860–1922) 189
Lessona, Carlo 179
Lessona, Michele (1823–1894) 4, 179
Levi, Beppo (1875–1961) 123
Levi, Donato (?–1885) 3, 178, 184
Levi-Civita, Tullio (1873–1941) 169
Lévy, Maurice (1838–1910) 56
L'Hospital, G. F. A. (1661–1704) 65
Liard, L. 46
Lie, Sophus (1842–1899) 110

Lindemann, Carl Louis Ferdinand (1852–1939) 103, 162
Lipschitz, Rudolf (1832–1903) 17
Liptay 127
Lobachevskii, N. I. [Лобачевский, Н. И.]¹(1792–1856) 71, 176
Loria, Gino (1862–1954) 11, 18, 65, 108, 130
Losio, Carlo 3
Lott, Julius 127
Lugli, Aurelio (1853–1896) 78
Lüroth, Jacob (1844–1910) 69

Maccaferri, Eugenio (1870–1953) 187
McColl, Hugh (1837–1909) 46
Macfarlane, Alexander (1851–1913) 46, 96
Maggi, Gian Antonio (1856–1937) 98
Maglioli, Fortunato 7, 177
Mago, Vincenzo 131, 142, 187, 191
Mansion, Paul (1844–1919) 13–14
Marcolongo, Roberto (1862–1943) 23, 86, 88, 122
Marconi, Guglielmo (1874–1937) 169
Marey, Etienne Jules (1830–1904) 56
Margherita di Savoia (1851–1926) 55
Marich, August 141
Marseguerra, Vincenzo 193
Martini, Eligio (?–1896) 4, 182
Marx, Karl (1818–1883) 138–139
Mastropaolo, Nicola 163
Mattirolo, Luigi (1838–1904) 169
Mattirolo, Oreste (1856–1947) 54
Mazzantini, Alessandro 146
Medvedev, F. A. [Медведев, Ф. А.] (1924–) 15, 19
Méray, Charles (1835–1911) 98, 108, 127
Meschkowski, Herbert (1909–) 70
Meyer, Friedrich 63
Meysmans, Jules 142
Michaux, A. 120, 128
Mie, Gustav (1868–1957) 33
Miracca, Raimondo 3
Mises, Richard von (1883–1953) 141
Mittag-Leffler, Gösta (1846–1927) 30, 83, 130–131

Möbius, August Ferdinand (1790–1868) 18, 22
Mohr, Georg (1640–1697) 166
Montessus de Ballore, Robert de (1870–1937) 167
Moore, Eliakim Hastings (1862–1932) 138
Moore, Gerald A. 142, 147
Mori Breda, Gilda 187, 192
Morrell, Ottoline (1873–1938) 138
Morris, Alice Vanderbilt (1874–1950) 162, 166, 169
Morris, Dave Hennen (1872–1944) 162
Mussolini, Benito (1883–1945) 163

Naccari, Andrea 29, 183
Nagy, Albino (1866–1901) 40, 46
Nassò, Marco 187
Natucci, Alpinolo (1883–1975) 157
Naville, A. 111
Netto, Eugen (1848–1919) 31–32
Newcomb, Simon (1835–1909) 58
Newton, Isaac (1643–1727) 26, 156
Nicoletti, Onorato (1872–1929) 188
Noether, Max (1844–1921) 185
Novarese, Enrico (1858–1892) 8, 40

Ohlhausen, G. R. 70
Ors y Rovira, Eugenio d' (1882–1954) 137
Osgood, William Fogg (1864–1943) 70

Padoa, Alessandro (1886–1937) 68, 78, 80, 83–84, 86, 89, 91, 93–98, 102–103, 108, 123, 130, 139, 166–167, 187, 189–191
Padula, Fortunato (1815–1881) 176
Pagliero, Giuliano 113, 122, 130, 141–142, 153, 187, 190
Pankhurst, Estelle Sylvia (1882–1960) 163, 169
Parfent'ev, N. N. [Парфентьев, Н. Н.] (1877–1943) 167
Parona, Carlo Fabrizio (1855–1939) 169–170
Parseval, M. A. (1755–1836) 54
Pascal, Ernesto (1865–1940) 189

Pasch, Moritz (1843–1930) 16, 27, 51
Pastore, Annibale, 190
Peano, Alessio (nephew) 1
Peano, Bartolomeo (?–1888) (father) 1
Peano, Bartolomeo (brother) 1–2, 145
Peano, Carmelo (nephew) 1
Peano, Carola (niece) 1
Peano, Caterina (niece) 1
Peano, Francesco (brother) 1
Peano, Giuseppe (1858–1932) passim
Peano, Giuseppina (niece) 1
Peano, Maria (niece) 1
Peano, Michele (brother) 1
Peano, Michele (1878–1942) (nephew) 1
Peano, Rosa (sister) 1
Peano, Rosa Cavallo (?–1910) (mother) 1, 131
Peirce, Charles Sanders (1839–1914) 21, 46, 73, 84
Pellegrino, Giuseppina 1
Pensa, Angelo 187, 190–191
Perna, A. 165
Peteri, Ilario Filiberto 3
Petiva, Federico 154
Peverone, G. Francesco 164
Peyroleri, Margherita 187, 190
Picard, Emile (1856–1941) 18, 41, 72, 78
Picone, Mauro (1885–1977) 167
Pieri, Mario (1860–1913) 25, 35, 45, 84, 88–89, 91, 94, 97, 105, 110, 112, 142, 187–190
Pierpont, James (1866–1932) 70
Pincherle, Salvatore (1853–1936) 69, 88, 165, 167
Pinth, J. B. 142
Pintor 169
Pivano, Silvio (1880–1963) 157, 167, 169
Plana, Giovanni (1781–1864) 153–154, 181–182
Plato (427 B.C.–348/347 B.C.) 82
Plücker, Julius (1801–1868) 176
Poincaré, Henri (1854–1912) 70–71, 93, 98, 110, 119–120, 139
Poli, Gino 191

Poretskii, P. S. [Порецкий, П. С.] (1846–1907) 46
Porro, Francesco (1861–1937) 36
Porta, Francesco 36
Posse, K. A. [Поссе, К. А.] (1847–1928) 15
Premoli, Orazio 7, 130
Pringsheim, Alfred (1850–1941) 14
Promis, Carlo (1808–1872) 177–178
Puccini, Giacomo (1858–1924) 42, 67
Puini, Carlo (1839–1924) 87, 90

Quarra, Paolina 140–141, 187, 191–192

Ramorino, Angelo 67, 102, 187–188
Rebstein 69
Resal, H. 58
Ricci, Umberto 130
Richard, Jules (1862–1956) 119–120
Richeri, Ludovico 52
Riemann, Bernhard (1826–1866) 176, 181
Rimondini, Filippo 189
Rosenberger, Woldemar K. 126
Royce, Josiah (1855–1916) 92, 123
Rudio, Ferdinand (1856–1929) 69
Ruggles, Charles (1892–1970) 168
Russell, Bertrand (1872–1970) 26, 73, 86, 91–92, 94–95, 99, 105, 110–112, 118–120, 122, 132–136, 138, 141–142, 148, 169, 174

Sannia, Achille (1823–1892) 176
Sannia, Gustavo (1875–1930) 134, 139, 146, 191
Sansone, Giovanni (1888–1979) 170
Santangelo, G. 190
Saussure, René de (1868–1943) 70
Schepp, Adolf (1837–1905) 15, 35
Schiaparelli, Giovanni (1835–1910) 61
Schiaparelli, Luigi 3
Schiff, Hugo Josef (1834–1915) 3, 178–179
Schiff, Moritz (1823–1896) 178
Schleyer, Johann Martin (1831–1912) 125–126, 128, 139, 166

Schoenflies, Arthur Moritz (1853–1928) 71
Schow 162
Schröder, Ernst (1841–1902) 21, 36, 38, 46, 71, 73, 78, 92, 94, 98
Schur, Friedrich H. (1856–1932) 71
Schurman, Jacob Gould (1854–1942) 92, 96
Schwarz, Hermann Amandus (1843–1921) 9–10, 18, 31, 33, 43, 78
Scott, Charlotte Angus (1858–1931) 70, 96–98
Segre, Corrado (1863–1924) 11, 35, 38–39, 71, 78, 80, 112, 144, 177
Serret, Joseph Alfred (1819–1885) 9
Severi, Francesco (1879–1961) 123, 170
Sheffer, Henry Maurice (1883–1964) 134
Siacci, Francesco (1839–1907) 6, 11, 42–43, 186
Sibirani, Filippo (1880–1957) 187, 192
Signorelli, Giuseppe 3
Silva, Giovanni (1882–1957) 159
Sineokov, N. S. [Синеоков, Н. С.] 15
Siriati, L. 161
Skryabin, Aleksandr [Скрябин, А.Н.] (1872–1915) 111
Smith, David Eugene (1860–1944) 43, 166
Somigliana, Carlo (1860–1955) 124, 139, 159, 169–171
Spezia, Giorgio (1842–1911) 4, 180
Staudinger, Franz 123
Staudt, Christian von (1798–1867) 35, 88
Stein, Ludwig 112
Steiner, Jacob (1796–1863) 72
Sterponi, Berardo 149–150
Stodola, Aurel Boleslav (1859–1942) 70
Stolz, Otto (1842–1905) 102–103
Study, Eduard (1862–1930) 110
Szilágyi, Dénes 141, 166

Tannery, Jules (1848–1910) 93–94, 111

Tannery, Paul (1843–1904) 111
Tanturri, Alberto 187, 191–192
Tardy, Placido (1816–1914) 13–14
Tarski, Alfred (1902–) 95
Teixeira, Francisco Gomes (1851–1933) 21
Terracini, Alessandro (1889–1968) 29, 131–132, 156, 170
Terracini, Ida 29, 41
Tessari, Domenico (1837–1909) 83
Thaon di Revel, Paolo Ignazio 169
Thomson, William (Baron Kelvin) (1824–1907) 58
Tibone, Domenico 6, 67
Timpanaro, Sebastiano (1888–1949) 171
Togliatti, Eugenio G. (1890–1977) 156
Tonelli, Alberto (1849–1921) 29
Tonelli, Leonida (1885–1946) 191
Torricelli, Evangelista (1608–1647) 101
Toscano, S. A. 191
Tricomi, Francesco Giacomo (1897–1978) 101, 156–157, 161, 167, 169–170, 172, 182
Turati, Filippo (1857–1932) 87

Umberto I (1844–1900) 55

Vacca, Giovanni (1872–1953) 71, 78, 80, 83–84, 86–88, 90–91, 101–105, 107–109, 111–112, 130, 134, 149, 166, 187, 191, 193
Vaccaro, Antonio 189
Vailati, Giovanni (1863–1909) 7, 25, 42, 44–45, 47, 49, 53, 67, 83–85, 87, 91–92, 96, 103, 109, 111–112, 123, 130, 187
Vasil'ev, A. V. [Васильев, А. В.] (1853–1929) 98, 110
Venn, John (1834–1923) 46
Veronese, Giuseppe (1854–1917) 39, 42, 51, 62
Vesin, Virginia 187
Viète, François (1540–1603) 49
Viglezio, Elisa 153, 156, 161, 187, 192
Viriglio, Luisa 187, 191–192
Vitali, Giuseppe (1875–1932) 189

Vivanti, Giulio (1859–1949) 18–19, 46–47, 62, 83, 187
Vocca, Paolo 193
Volterra, Vito (1860–1940) 15, 17, 42–43, 53, 55, 57–61, 64, 71, 84, 90, 130, 159

Ward James (1843–1925) 73
Weber, Heinrich (1842–1913) 69
Weierstrass, Karl (1815–1897) 102, 181
Whitehead, Alfred North (1861–1947) 99, 134–136, 141
Widmann, John 49
Wiener, Norbert (1894–1964) 69, 94, 134

Wilson, Woodrow (1856–1924) 151

Young, W. H. (1863–1942) 19

Zama, Mario 127
Zamenhof, Ludovic Lazarus (1859–1917) 98, 108, 120, 128
Zanotti-Bianco, Ottavio (1852–1932) 58
Zavagna, Ireneo 187, 192
Zeni, Tancredi 10
Zermelo, Ernst (1871–1953) 33, 118–119, 122, 138
Ziwet, Alexander (1853–1928) 43

INDEX OF SUBJECTS

Academia pro Interlingua 130, 133–134, 139, 146–147, 162

Accademia dei Lincei, Peano elected to 112, 166

Actuarial questions, Peano's treatment of 102, 113

Agnesi, curve of 65

Axiomatics 27, 118–119

Axiom of choice 33, 104, 119

Burali-Forti's antinomy 118

Calculus, geometrical 21–23, 35, 57, 64–65

Calendar 155, 159, 161, 163–164, 166

Cantor-Bernstein Theorem 118

Cavoretto, workers' outing to 115–116

Curve, space-filling, Peano's 31–32, 67, 173

Definition by abstraction 46, 82, 148

Definitions in mathematics 77, 93–94, 104, 136, 154

Definitions, recursive 174

Délégation pour l'adoption d'une langue auxiliaire internationale 112, 121, 123, 128, 144

Derivative, definition of 41

Derivatives: mean-value theorem for 15–16; commutativity of second partial 14, 32–33

Differentials and derivatives 138–139

Differential equations: method of successive approximations 17–18, 78, 174; Peano's theorem for 17, 33, 173–174; systems of 41

Earth's pole, motion of 57–61

Ellipse, formula for perimeter of 30–31

Ellipsoid, area of 35, 40

Esperanto 96, 98, 108, 120, 127–128, 143, 154

Euclid's *Elements* 32, 36, 40, 101, 120, 176

Fermat's Theorem 103, 162

Formulario project 44–50, 53, 64–65, 76–77, 121, 175

Functions: complex, interpolating 11; discontinuous in every interval 40; elliptic, geometrical definition of 21; nowhere differentiable 32; of sets 18–19, 174

Geometry: Peano's axioms for 27, 51–52; Pieri's axioms for 88, 112

Grammar, algebra of 136, 167

Ido 128, 144

Infinitesimal segments, impossibility of 40

Infinitesimals, Peano's views on 131

Integral, Peano's definition of 11, 15, 174

Interlingua 107, 133–134, 136, 154, 160, 166, 175

International Auxiliary Language Association (IAIA) 154, 162, 169–170

Latino sine flexione 105, 107–111, 122, 125, 129–130, 136, 138

Limit of a function, Peano's definition of 40–41, 52, 64

Lobachevskii Prize 110

Logic, mathematical: Frege's notation for 74–76; Peano's 21–22, 26, 28, 36, 46–49, 52, 67–69, 74–76, 141–142, 148, 174–175

Mathematicians, International Congress

of: Bologna (1928) 165; Cambridge (1912) 138; Paris (1900) 96–98; Rome (1908) 122; Strasbourg (1920) 159; Toronto (1924) 159–160; Zurich (1897) 69–72
Mathesis, società italiana di matematici 68, 78, 80, 82, 103, 123, 130, 140, 145, 152, 157, 161
Measure, Jordan-Peano 18, 174
Military Academy (Turin) 17, 30

Numbers: cardinal, Cantor's definition of 62; irrational 82; natural, axioms for 25–27, 37–38, 80, 88, 174
Numerical approximations 148–152

Padoa's Method 95
Peano, Giuseppe: as student 3–6; as teacher 7, 43, 100–102, 131–132, 137; as typesetter 80, 104, 148; as university assistant 7, 10; childhood of 1–2; competes for chair of infinitesimal calculus 29–30; death of 168; elected member of the Academy of Sciences of Turin 35; funeral of 169; honors for 55, 100, 150, 155; illnesses of 34, 63, 112; introduces decimal ordering of propositions 83; linguistic ability of 38, 46; marriage of 20; named professor at the University of Turin 35; political views of 114, 147, 173; professor at the Royal Military Academy 17; publication of the *Opere Scelte* of 170–177; purchases land 42, 104; purchases a printing press 79; purchases villa in Cavoretto 40; receives *privata docenza* 16; relatives of 1–2; religion of 2; residences of 7, 10, 16, 20, 40, 67; style of living of 67; views on teaching of 131, 137, 140, 158
Philosophy, International Congress of: Bologna (1911) 134, 136–137; Geneva (1904) 111–112; Heidelberg (1908) 122–123; Naples (1924)

159–160; Paris (1900) 92–96

Quadrature formulas 18, 42, 140–142, 145
Quaternions, International Association for Promoting the Study of 23
Quantification: Peano's symbols for 140; universal 26

Relativity: Burali-Forti's views on 86, 160; Peano's views on 160
Revista Universale 142
Richard's Paradox 119–120
Rivista di Matematica, founding of 35–36
Ro 148
Russell, Bertrand, Peano's influence on 91–92, 94–95, 132–133, 138, 169, 174

Schola et Vita 163
School of Peano: Boggio 89–90; Burali-Forti 85–86; Castellano 85; Padoa 86, 139; Pieri 88–89; Vacca 86–88; Vailati 84–85
Set theory 21–22, 26, 33, 46
Simpson's Rule 41, 145
Surface, curved, definition of 9–10

Taylor's Formula 14, 28, 39, 51
Turin, University of: foundation of 109; student disorders at 8–9, 16–17, 20–21, 25, 40, 51, 54–55, 66–68, 109, 146; women students at 29, 122, 132

Vector analysis, Italian 23
Vector space, axioms for 23–24, 174
Vector theory 79, 104–105
Vectorial methods, opposition to 85
Volapük 108, 125–129, 139, 166
Volapük Academy 126–127, 129–130, 142

Wronskians 28, 69

STUDIES IN THE HISTORY
OF MODERN SCIENCE

Editors:

ROBERT S. COHEN (Boston University)
ERWIN N. HIEBERT (Harvard University)
EVERETT I. MENDELSOHN (Harvard University)

1. F. Gregory, *Scientific Materialism in Nineteenth Century Germany.* 1977, XXIII + 279 pp.
2. E. Manier, *The Young Darwin and His Cultural Circle.* 1978, XII + 242 pp.
3. G. Canguilhem, *On the Normal and the Pathological.* 1978, XXIV + 208 pp.